for label

used in flora lists

I = small pieces take root easily
J = rosette habit
K = self-sown garden plants
L = found in lawns, though not necessarily confined to lawns
M = found on paths, though not necessarily confined to paths
N = seeds self-dispersed
O = seeds dispersed by animals
p = creeping or prostrate habit
P = seeds dispersed by wind
Q = early colonizer
R = perennates by rhizomes
Ro = perennates by thick root system
S = perennates by stem stock or root stock
Sap = saprophyte
T = perennates by tubers
Tap = perennates by long tap root
Tur = reproduces by turions
V = insectivorous plant
W = woody perennial (tree, shrub or woody climber)
X = xeromorphic plant
Y = halophytic plant
Z = most persistent garden weed.

Note. The categories listed above are not applied to all lists, but only where necessary to bring out some point dealt with in the text. For example, the *absence* of 'O' beside *Anthriscus sylvestris* in the hedgerow list (p. 91) does *not* imply that its seeds are not dispersed by animals. The subject of seed dispersal method is discussed only in connexion with waste land, and it is only in the waste-land list that categories of seed dispersal method are indicated.

Mosses

CM = cushion moss
FM = feather moss

Lichens

BL = branched lichen
CL = crusty lichen
LL = leafy lichen

Algae

BGA = blue-green alga
FA = filamentous alga
PP = phytoplankton (non-filamentous, single-celled or colonial algae, floating or swimming free in the water.

Liverworts

LLW = leafy liverwort
THL = thalloid liverwort

Fungi

BF = bracket fungus
CF = cap fungus

NATURAL COMMUNITIES

O. N. Bishop

John Murray London

Printed in Great Britain by Martin's of Berwick

0 7195 2246 3

Preface

This book provides a concise survey of the chief types of natural community found in Great Britain. It should be of interest to naturalists, and to sixth-form and undergraduate biologists, as well as to the audience for which it was originally conceived—the teachers using my series, *Outdoor Biology*, in middle and secondary schools. The book is intended not just for arm-chair reading but as a preliminary to some sort of practical field study, however slight. It provides a basis from which the reader can learn to analyse natural communities and in doing so discover far more than is in this book.

Wiseton ONB

Contents

Colour plates

INTRODUCTION

A natural community consists of a number of different species of animals and plants living together in a particular locality. Though no two localities are inhabited by exactly the same collection of species, there are general patterns of community membership that allow us to recognize distinctive communities such as a wood, a heath, or the community of organisms living on a sand dune. This book explains in outline the nature and composition of the natural communities most commonly found in Great Britain.

The British countryside and the urban areas display a very wide range of communities and it is not possible to make a really detailed study of any of these in a book as short as this. For detailed study the reader is referred to the book lists (pp. 175-8).

For the same reason the reader must not expect to go to a wood and find that the plants there correspond exactly to those listed in this book. Each community is unique because it depends on a unique combination of local conditions. The lists in this book are a guide to what is *most likely* to be found. Some listed plants may be absent from a given location because local conditions do not suit them (here is scope for investigation) and, conversely, a plant that is uncommon over Britain as a whole, and thus not listed, may be very common in the given location. Rarities do not appear in the lists, even though they may be described by some authors as characteristic of a community, for rarities—because of their rareness—have little ecological influence.

One point requires special emphasis. This is a book about communities, and there is not enough space to deal with individual species or plants or of animals. To understand the nature of a community one needs to get to know some of the individual species of which it is composed. This is best done by field studies. The flora and fauna lists of this book tell the reader what to expect in the field, but they do not give descriptions by which the species may be recognized. It is important to acquire some of the key works and perhaps some of the descriptive works listed on pp. 176-8, and to use these in conjunction with this book when working in the field or when making preliminary studies at home.

It has not been possible to prepare comprehensive fauna lists for each community. One difficulty is that there are so many animal species that any list of publishable length would include only a fraction of those likely to be found. A further point is that many common animals are quite unlike plants in the way they are distributed over the countryside. Plants tend to occur in recognizable communities, but animals more often occupy microhabitats within these communities and occur wherever these microhabitats exist. Woodlice will be found wherever there are suitable dark damp conditions, whether it be underneath a stone in a wood, beneath a stone on a hedgerow verge, or under a stone on a derelict city site. Hedge sparrows will be found wherever there is a convenient nest-site. Slugs are found wherever there is damp soil with abundant vegetation, and the holly-leaf miner wherever there are holly trees. It remains true that many animals do have a distribution based on geographical features or on habitats (for example, the freshwater animals), but the tendency to distribution by microhabitats, in conjunction with the large number of species and the difficulties of identification of small animals, make fauna lists unprofitable. Instead, an attempt has been made to analyse each habitat into its microhabitats and to suggest very common examples of species likely to be found inhabiting them. The species actually found may be quite different from those suggested in this book, but being in the same or a similar microhabitat they will be more or less identical ecologically.

There is a deliberate bias in the regions selected. Attention has been given chiefly to those areas most likely to be of concern to school ecologists (woods and freshwater aquatic habitats) or to be most accessible (hedgerows, parks, gardens, walls, waste ground). Other regions, equally interesting and important from a purely ecological view, have been dealt with only briefly on the grounds of their lesser interest for school ecologists or their lesser accessibility to most schools.

WOODLAND

Kinds of woodland

In most woodlands one or two species of trees outnumber the other species present. These are the dominant species, or co-dominant species when two species are present in more or less equal numbers, and it is by these that different types of woodland may be distinquished. Being present in the largest numbers, and casting shade on all plants below, the dominant trees exert a considerable influence upon all other plant life there. In its turn, the plant life influences the animal life of the wood, hence the dominant tree species determines to a very large extent the pattern of life for the whole wood.

This pattern will be analysed for a number of kinds of woodland commonly found in Britain:

1. Damp oakwood
2. Dry oakwood
3. Beechwood
4. Ashwood
5. Birchwood
6. Pinewood
7. Alder carr

These are fundamental woodland types which may be modified by local conditions. Certain combinations of soil and local climate may produce a pattern differing in several ways from the fundamental types. Many woods have been planted with mixtures of trees that do not conform to the fundamental types, so that this *mixed woodland* may show intermediate features. Man may modify the patterns in woods, either by cultivation, by local pollution, or by many types of use and abuse.

The first woodland to be described is *damp oakwood.* In that section a number of features common to all woods will be considered. Descriptions of other types of wood will be confined mainly to points of difference.

Damp oakwood

This may be regarded as the typical English woodland, and it is believed that most of the country was at one time covered with woodland of this kind. There is probably no woodland in Britain today that has not been influenced at some time by man's activities, so that practically all oakwoods now show evidence of modification from their natural state. Damp oakwoods are found growing on heavy loam or clay soils. They occur mainly in southern, central, and eastern England. By contrast, the dry oakwoods, to be described later (p. 24), are found on shallow, lighter, sandy soils, and occur mainly in the west of Britain, on the old rocks of south-west England, in Wales, the Pennines, and Scotland. They are also found on the gravels and sands of Hertfordshire, Surrey, and Kent.

The dominant tree of damp oakwood is *Quercus robur*, the common oak, sometimes known as the pedunculate oak because of the long stalks or peduncles by which the acorn cups are attached to the twig. Growing among the oaks may be a few individuals of other species of large tree, all reaching heights of about 30 m, like the oak. These are listed on p. 17. Trees that reach lesser heights (15 to 25 m) may be found around the edges of oakwoods, or in open spaces within the wood. These trees may have been planted by man, or their seeds may have been carried by wind or animals and managed to establish themselves. At the wood margins, in open spaces, or where the continuous cover of oaks has been breached by felling or the death of an oak tree, these seedling trees have the best chance of becoming established.

Which of these species will be present in a given wood depends partly on the previous history of the wood. It is unlikely that all would be found within one wood, and it is possible that other species not listed here may have been planted. The environment of the wood may favour one species especially; for instance, *Alnus glutinosa* is more usually found in woods on wet soil. Trees planted less commonly may include *Quercus cerris*, the Turkey oak, with bristly stipules at the base of its buds and bristly cup scales, and *Acer campestre*, the field maple. This is sometimes mistaken for *Acer pseudoplatanus*, the sycamore, but may be distinguished by its smaller size, by the soft short hairs on its twigs (sycamore has no hairs), and by the wings of its paired fruits which are nearly 180° apart instead of about 120° as in sycamore.

Two other species of lime may be found planted in woods, but the one listed is the commonest, and is a hybrid of the other two species. This is indicated by the '×' before its specific name. This species is a hybrid between *Tilia cordata*, the small-leaved lime, and *T. platyphyllos*, the large-leaved lime.

In a wood the oaks, together with other trees of similar stature, extend their crowns to make a more or less continuous canopy. Below this the environmental conditions differ from those outside the wood in several ways:

1. *Light intensity*. This is reduced by shade cast by the leaves, and only plants able to grow under reduced light intensities can survive within the wood. Others are confined to margins and open spaces where the light intensity is greater. The shading effect in oakwoods is seasonal, for the trees are leafless from mid-September until mid-May.

2. *Temperature*. The crowns protect the ground below from the direct radiation of the sun, so reducing daytime temperatures. Also, they reduce

radiation of heat at night, so maintaining a relatively high temperature at night. The effect of the trees is thus to moderate the extremes of temperature, and create a more equable environment.

3. *Wind.* The trees give protection from excessive wind force, and, in general, reduce wind velocities.

4. *Humidity.* With reduced wind velocities the water vapour transpired from the plants is not so readily carried away from the wood. In addition, the lower daytime temperatures in the wood reduce the water-holding capacity of the air. For these two reasons relative humidity in the wood tends to be higher than that outside, and this factor, taken with the reduction in temperature and wind velocity, lowers transpiration rates of plants in the wood.

5. *Soil condition.* A certain fraction of the rain falling on the canopy remains on the leaves and branches and evaporates directly from there. This reduces the amount of water available for the plants below, but the effect is slight and probably more than balanced by their low transpiration rates. Plants need not suffer water shortage on this account. The primary effect of the trees on soil water content derives from the uptake of water through their roots. In damp oakwoods the soils are usually deep and the trees deeply rooted, leaving the water in the surface layers for use by the other plants (but see p. 25).

The fall of leaves in autumn deposits a large quantity of organic material on the soil, and in time this rots to form rich humus in the surface layers. This gives the soil good physical properties and an ample supply of mineral ions. In some types of woodland this does not happen (p. 41).

Stratification

When studying the flora of woodland it is usual to consider the plants as if they were arranged in several distinct layers. This is convenient for discussion, and enables the flora list to be arranged in manageable sections. In a real wood these layers are usually less distinct than they appear on paper; the layers interact in a complex way and can rarely be considered in isolation. There are usually five layers:

1. *Tree layer*

This comprises the dominant and other equally tall trees. The characteristics and influence of this layer have already been dealt with.

2. *Second tree layer*

This consists of trees reaching heights between 5 and 15 m whose crowns come below or partly below those of the trees in the tree layer proper. These lower trees are shaded by those above, so only those able to live in reduced light intensity can form part of this layer. Some show special features that help them to survive. For example, the evergreen trees *Ilex aquifolium* (holly) and *Taxus baccata* (yew) have leaves throughout the year, which compensates for their being shaded during the summer. Their leaves contain a relatively large proportion of chlorophyll, which probably makes it possible for them to absorb light more efficiently. *Ilex* may sometimes become the dominant tree of this layer, while on chalk or limestone soils the dominant tree is often *Taxus*. Usually, the trees are not numerous and are scattered singly in the wood. If either of them is particularly abundant then they may form a second canopy beneath that of the upper tree layer. Beneath this double canopy, light intensity is extremely low, and even in winter there is still the evergreen canopy, so there is little plant life below. Beneath single trees of *Ilex* or *Taxus* the soil is usually bare. In open woods (woods in which the dominant trees are widely spaced) the second tree layer may also contain individual young trees that will one day reach maturity and occupy their place in the upper tree layer.

3. *Shrub layer*

This may be difficult to distinguish from the second tree layer in practice, because the shrubs also are woody perennials, and some shrub-like plants may be on their way towards growing large enough to be reckoned among the second tree layer. *Corylus avellana* (hazel) is often the dominant plant of the shrub layer, and *Crataegus monogyna* (hawthorn) may commonly be found there. Where rabbits are common, *Sàmbucus nigra* (elder) may form dense thickets. Rabbits are generally very destructive to seedlings of trees and shrubs, but elder is resistant to their attacks, possibly because of the unpleasant smell of the leaves and stems when they are crushed. *Rubus fruticosus* (bramble) frequently forms a dense undergrowth which partly merges with the field layer below. Its branches spread out from the base of the shrub, arch over, and touch the ground with their tips. Here they take root, forming new plants still attached to the parent plant by the arching branches. In this way *Rubus* produces a loose tangle of branches within 0·5 m of the ground, between which other plants can grow. *Carpinus betulus* (hornbeam) is included in the shrub layer list even though it can grow to a tree up to 30 m high in the open. Usually in a wood it is coppiced (p. 12) and so remains with a shrubby habit.

One point of interest in the tree and shrub layers is that a high proportion of the species produce their flowers before they produce their leaves (p. 17), where FBL = flowers before leaves, and FWL = flowers with leaves). A possible advantage of this is that flowering and fruiting are well advanced before the leaves begin to expand and partly shelter them from the wind. These species all rely on the wind either for pollination or fruit dispersal or both. In woodland this is advantageous, for flowering and fruiting will be over before the canopy of oak leaves has become complete.

Four species of *Salix* (willow) have been listed, but other species may be found instead of these, and frequently a specimen seems to fit none of the descriptions exactly. *Salix* often hybridizes, and where two or more different species are found in a locality, hybrids between them may be quite common.

Included among the shrub layer are three woody climbers. *Hedera helix* (ivy) may spread over the soil forming a loose mat among which other species can grow. In this way it is frequently part of the field layer. It can also climb the trunks of trees, attaching itself by its adventitious roots, reaching a height of up to 30 m, and so extending upwards into the tree layers. *Hedera* has dark green leaves containing much chlorophyll, which presumably adapts it to low light intensities. The other two climbers listed have lighter green leaves, and are found only on the margins of woods and in open spaces. Of these *Lonicera periclymenum* (honeysuckle) can climb to only 6 m, while *Clematis vitalba* (old man's beard) can reach 30 m. *Clematis* is common only on calcareous soils in the south of Britain.

4. *Field layer*

This consists of non-woody plants, usually considerably less than 1 m high, the only exception in the lists being the small woody plant *Solanum dulcamara* (woody nightshade). They live beneath the shade of the tree layers and the shrub layer, and experience even lower light intensity, slighter winds, more equable temperatures, and higher humidity than do the woody plants of the upper three layers. With this protection from the extremes of the elements, a woody habit and a bark-covered stem are little advantage. This protection has its attendant disadvantage in that it is difficult for a seedling plant to establish itself under such conditions. A young seedling has only the food reserves present in its seed, and when it germinates in spring these must be used to make the first leaves and roots. From then on, when its reserves are exhausted, it depends on its own photosynthetic powers to provide materials for continued growth. Under woodland conditions this is the time at which the canopy above is closing and light intensity is being drastically reduced. Under these conditions it is difficult for a seedling plant to become established, and those plants that need to begin each year by germination from a seed are at a disadvantage. Consequently few annual plants are found in woodland (only three are listed) and few biennials either (only three listed) and these are confined to the margins or open spaces of the wood. By contrast, the perennial habit is a great advantage in woodland for, with ready-formed buds and a large reserve of food, a plant can produce its leaves early in the year, before the oak canopy has closed over. Growth is restricted then, and the remainder of the season can be spent in accumulating reserves ready for an early start in the following spring. Hence many woodland plants (though not all) show an early spurt of growth and flowering, and are less active or may even die back during summer. The best examples of these plants are listed under 'Early spring flowers' on p. 18. They include many species with bulbs, corms, tubers, and rhizomes, for these are perennating structures. As well as these, many woods contain early flowering plants that have been planted there or escaped from cultivation in parks or gardens nearby. These could include *Narcissus pseudonarcissus* (daffodil), *Tulipa* spp. (tulips, including *Tulipa sylvestris*, the wild tulip) and *Galanthus nivalis* (snowdrop). The flowering period of these early plants begins in December in sheltered conditions (for *Primula*) with the majority flowering from March to April.

Following these come those that flower from April to May or June, so that the damp oakwood is characterized by its continuous and varied display of spring flowers, which makes it an attractive place, and makes spring an encouraging time to begin field studies. In the summer months other plants come into flower, though most are not so showy as the spring flowers and many show adaptation to growing in shade conditions. They may have broad or dark-green leaves (*Allium ursinum*, *Stachys sylvatica*, *Urtica dioica*), or need the cool humid conditions found under the summer canopy. For example, *Oxalis acetosella* (wood sorrel) has leaves of three leaflets which fold together at night and also in direct sunshine. This is thought to be a mechanism for reducing transpiration. Ferns may commonly be found in the shadier parts of the wood. They seem to grow best under these conditions, and at the stage in their life-history when the male gamete swims in a water-film to fertilize the female gamete, damp conditions are absolutely essential.

The length of the list of field layer plants indicates that most damp oakwoods contain a rich flora; but all these would not necessarily be found in a single wood. The lists include only widespread species, but others may be abundant locally. Much depends on local conditions; for example, in low-lying or badly drained soils where soil moisture is high on average, those plants best suited to wet soils would be expected to be commonest. Of the plants listed, those associated with wet soils are:

Allium ursinum
Angelica sylvestris
Caltha palustris
Carex spp.
Chrysosplenium angustifolium
Cirsium palustre
Deschampsia caespitosa
Epilobium adnatum
Filipendula ulmaria
Galium odoratum
Juncus spp.
Listera ovata
Lychnis flos-cuculi
Myosotis scorpioides
Ranunculus repens
Scrophularia spp.

If the woodland was very swampy, the above might be found, and also *Iris pseudacorus*. Like several other species listed, *Iris* is not regarded as a typical woodland plant, for it is more commonly found elsewhere—on the margins of ponds and streams, for example. Likewise, many plants more typical of meadows and hedgerows may become established, temporarily at least, on the margins of woods or in open spaces or clearings. Among the listed species, the following are found on margins, in openings, or throughout open oakwoods:

Alchemilla vulgaris
Anthriscus sylvestris
Arctium minus
Betonica officinalis
Chamaenerion angustifolium
Cirsium palustre
Epilobium adnatum
Galium mollugo
Geranium robertianum
Heracleum sphondylium
Hypericum perforatum
Lapsana communis
Prunella vulgaris
Tamus communis
Torilis japonica
Veronica officinalis

This includes all the annual and biennial plants from the woodland flora list.

If the vegetation can vary from one wood to another, it can also vary within a wood according to small-scale variations in environmental conditions: differences of drainage or proximity to a pond or stream may make parts of a wood damp, so encouraging some of those plants associated with wet soils, and differences of light and shade may influence the distribution of others. Effects of this kind provide frequent scope for study in the field. Some plants, such as the typical hedgerow and meadow plants listed above, favour the better-lit regions, while others establish themselves in the deeper shade. In shady spots one may find the evergreen shrubs, many ferns, *Urtica dioica* and *Stachys sylvatica*. Like *Sambucus* (p. 6), *Urtica* is favoured by a large local population of rabbits, which avoid it because of its stinging hairs. They devastate the other vegetation but leave the nettles untouched. Other plants avoided by rabbits are *Rubus fruticosus*, *Cirsium palustris* (and other thistles), *Glechoma hederacea* and *Teucrium scorodonia*. Apart from rabbits, other herbivores such as mice, snails and slugs, and the larvae of very many insects, eat the leaves, stems or roots of plants. Oak leaves are the food of many such herbivores, and the oak is the host of many parasites, including several gall-producing insects. Holly leaves are commonly mined by certain insect larvae, and leaf-miners are found on many other trees and field-layer plants. A heavy local infestation may have serious effects.

We must not forget man and his activities in and around the wood. Near to towns the leaves of trees and other plants become covered with sooty deposits, and the air contains fumes that restrict the growth of some species, especially the lichens. If a wood is open to the public and is near a large town it may become a favourite haunt for picnickers and others at weekends, with deleterious effects on the vegetation. These effects have not yet been studied extensively, but they are striking. Among the most serious are damage to trees, indiscriminate picking of wild flowers (especially the early-flowering species, which in some areas are becoming almost extinct), and the general trampling of vegetation. Few plants seem able to survive continued trampling, but an example of the type best able to withstand trampling is *Sagina procumbens* (procumbent pearlwort). This is common in woods at the edges of pathways. Like other pathway plants, it has a low, spreading habit. The effects of trampling make a good topic for *experimental* investigation.

Among the herbaceous plants of the field layer one may sometimes find a surprisingly large number of seedling trees, perhaps as much as 20 cm high. Yet in the shrub layer not nearly so many saplings are found, and a short calculation shows that only a small fraction of these would ever be required to fill gaps in the tree layer caused by the death or destruction of older trees. What becomes of these seedlings? This question is a starting point for a series of practical investigations. Its main drawback from the viewpoint of the school pupil is that such investigations may need to extend over several years.

5. *Ground layer*

In the better lit parts of a wood the individuals of the field layer are so closely packed that there is no space between them and no ground layer. In other parts of the wood the field layer plants may be more widely spaced, owing perhaps to a lower light intensity or difficulties of rooting in very shallow soil such as might be found around the base of a tree-trunk. In such situations, between and beneath the plants of the field layer, the plants of the ground layer can become established. These are all small and consist mainly of mosses, liverworts, and the prothalli of ferns. Owing to the low light intensity beneath four layers of vegetation they may be able to grow only slowly, but their prime need is for a cool, humid atmosphere, for these plants are easily desiccated in warm, dry situations. Although some mosses are able to resist desiccation very effectively, they are not able to grow or to reproduce while in the dried condition. For sexual reproduction, in which the male gamete must swim in a film of water, the mosses and liverworts, like the ferns, need moist conditions.

In waterlogged hollows and in the region of semi-permanent puddles the soil surface may become covered with a film of algae, particularly those known as blue-green algae.

Unless one is making a detailed study of mosses and liverworts it is usually not worth while to identify the species precisely. A simple division of liverworts into thalloid and leafy types and a division of mosses into cushion and feather types will suffice. The list shows that both are present in woodlands. If more detailed work is carried out, the effects of variation in environmental conditions can be investigated. For example, *Eurhynchium* is often found in heavy shade, *Mnium affine* and *Thamnidium* require wet conditions, and *Fissidens* and *Mnium punctatum* seem to require both wet and shady conditions.

Epiphytes

These are plants that grow on other plants, not taking nourishment from them or acting as parasites in any sense but merely using the plant as a convenient support. The most common support in a wood is the bark of the trees, which, being slightly absorbent, holds moisture, and because it is slightly fissured provides a hold for germinating spores and seeds. In the open the bark of trees is normally too dry for much epiphytic growth to occur, especially on windward or sunward sides of trunks and branches, but in the shady, humid atmosphere of a wood the bark remains damp for long enough to enable epiphytes to become established. In tropical forests many higher plants such as ferns and orchids grow epiphytically, but the only common larger epiphyte in Britain is *Polypodium vulgare*, a fern and also a member of the field layer. The other epiphytes, which may be prolific in growth, are all members of groups also found in the ground layer, including mosses, liverworts and algae, or are lichens.

Usually, the first epiphytes to appear on bark are the algae, of which the commonest is *Pleurococcus*. In many situations this may be the only epiphyte found. It forms a matt green layer on shaded parts of trunks and branches, and is also common on fences, rocks and walls. It consists of microscopic cells that are spherical in shape. The other common alga, by no means as common as *Pleurococcus*, is *Hormidium nitens*. This, too, is green, but is filamentous.

Lichens are well adapted for epiphytic existence, for they are resistant to extremes of desiccation, and are slow-growing so that they can exist with the minimum of mineral matter, such as they can obtain from wind-borne dust which is carried to the bark, and settles in crevices. They need light so that the algal partner is able to photosynthesize to produce organic materials. Unless one is making a special study it is seldom that exact identification is needed, though it is helpful to divide them roughly into three groups, according to whether the lichen has the form of a branching stem (BL), a leafy, flaky appearance (LL) or forms a flattened, crusty colony close to the surface (CL). Lichens do not grow well in proximity to towns. Sometimes the only common lichen on trees in or near towns is *Lecanora conizaeoides*. At the other extreme, in areas remote from towns, trees may be covered with branching and leafy lichens in profusion. Sometimes a dead tree, barren of leaves, looks from a distance as if it were in full leaf, such is the growth of lichens on its branches.

Mosses on the whole need slightly moister conditions and a rather greater accumulation of soil particles. *Orthotrichum diaphanum* is commonly found in town areas. The liverworts require moist conditions and are most often found nearer the ground in sheltered spots. On old decaying tree stumps the wood and bark become rotten and soggy, thus making ideal places for liverworts to grow.

Oakwoods are a good place for the study of epiphytes for, apart from the favourable conditions provided by the woodland habitat, the bark of oak trees is not as acid as that of many other trees and is therefore more favourable to the growth of lichens, algae and mosses.

Fungi

Dead plants, and particularly the leaves cast from the trees in autumn, fall to the soil where they accumulate as a layer of leaf litter. In the autumn, in damp oakwoods, the weather is still warm and often rainy and the soil is usually base-rich, giving ideal conditions for the growth of fungi which send their hyphae into the leaves and other plant remains, removing organic and inorganic substances for use as food. The rotting action of the fungi does continue at other times of year, and leaves may not finally be rotted until the following spring, but the autumn is the best time for finding the fruiting bodies of those fungi that produce ones large enough to see. The species of fungi that may be found in a wood are many, and their identification is often a matter of difficulty. The list (p. 23) gives only a few of the common ones. Those listed as cap-fungi are found on oak leaves in the autumn and early winter, except for *Mycaena inclinata* which grows on stumps of oak trees. The bracket fungi are found growing from the trunks of oak and other trees. Whereas the fungi on leaves are performing a useful service to the woodland community by helping to reduce the materials in the leaves to a form assimilable by plants, the bracket fungi are parasites, obtaining nourishment from the living oak tree, usually weakening or destroying its wood.

Coppicing

In woods tended by their owners with the object of eventually producing saleable timber, the young seedling trees and bushes are cut back every few years, so promoting the growth of the older oak trees. Coppicing, and the frequency with which it is practised, has a great effect on the vegetation within the wood. If it is left uncoppiced, the shrub layer becomes dense after a few years. Below this thicket there is seldom enough light to support more than a few of the most shade-tolerant species. The carpet of spring flowers and the profuse summer vegetation of a coppiced oakwood are absent.

At the other extreme, a wood that is frequently coppiced loses some of the characteristics of damp oakwood, for the continual opening up of the shrub layer allows increased light and reduces the average humidity. This favours the spread of those plants that do better in higher light intensities, such as *Ajuga reptans*, *Anemone nemorosa*, *Endymion non-scriptus*, *Galeobdolon luteum*, *Primula vulgaris*, and *Silene dioica*. A wood in which part is coppiced and part is left untended could make an interesting study.

Animals in oakwoods

The pattern of plant life in a wood is complex, and because of this a wide range of environmental conditions has been created. One would expect that many different species of animals could each find within the wood some conditions suited to their needs. In other words, there are many micro-habitats, and because of this many species of animal can live side by side in the wood, without competing greatly with one another, for they feed on different food-plants and live in different parts of the wood. This is the basis on which the pattern of animal life will be analysed. Accurate identification raises a slight problem here, for with many of the small animals this is difficult, time-consuming, and requires considerable expertise. It is preferable to content oneself with identifying only as far as the family or genus and to concentrate more attention on the ecological role of each type of animal collected. The aim should be to define its ecological niche. For a given specimen of insect it is more useful to say, 'This is a Capsid Bug we found feeding on the sap from oak leaves, using its sucking mouthparts' than simply to record its scientific name, *Lygocoris viridis*.

Animals differ from plants in being mobile; they can shelter in one place and feed in another. Most remain in the wood for both food and shelter, but some (for example, heron) shelter in the wood but feed elsewhere, and others (for example, pigeon) may enter the wood to feed, yet nest outside. The trees, shrubs and smaller plants provide food for animals, and since these plants are arranged in layers, the animals feeding on them are likewise stratified in distribution. Within each stratum certain animal species may be restricted to leaves, or to stems, or to other parts of the plant. The provision of food by the wood may be summarized like this:

TREES AND SHRUBS

Leaves	eaten by	larvae of moths (green tortrix on oak, oak roller on oak; oak eggar on bramble, hazel and ivy, but *not* on oak; buff-tip on elm, lime, hazel)
		larvae of sawflies (oak sawfly on oak or willow)
		adult cockchafer
	sap sucked by	aphids (oak leaf aphid)
	mined by	larvae of moths
		larvae of other insects (*Lithocolletis* sp. on oak, hazel and hawthorn)
Bark	eaten by	rabbit
		common vole
	bored by	larvae of weevils (oak bark beetle; goat moth in willow)

Fruits and seeds	eaten by	birds (robin, green woodpecker)
		mammals (mice, squirrels, voles)
	bored by	larvae of weevils (nut weevil in hazel nuts and acorns)
	sucked by	shieldbugs (hawthorn shieldbug)
Galls	Over 50 different insects cause galls on oak, among which are	gall wasp, *Biorrhiza pallida* (oak apples)
		gall wasp, *Cynips kollari* (oak marble galls)
		gall wasp, *Neurotus* sp. (spangle galls—these, too, may show stratification, because some species are commoner near the top of the tree, others are commoner near the middle or at the bottom of the crown)

FIELD AND GROUND LAYERS

Leaves and stems	eaten by	rabbit
		slugs (*Limax maximus*)
		snails (*Cepea* sp.)
		larvae of insects (snout on stinging nettle, oak eggar moth on various plants)
	sap sucked by	aphids
		capsid bugs (*Dicyphus errans* on stinging nettle, hedge woundwort, herb Robert)
	mined by	larvae of insects (celery fly in cow parsnip)
Roots	eaten by	larvae of insects (many, including crane-fly)
		roundworms
Fruits and seeds	eaten by	mice
		birds (hedge sparrow, yellowhammer)
		squirrels
	sucked by	insects (capsid bug on fern sporangia)
Pollen and/or nectar	taken by	butterflies
		bees
Leaf litter	(including that from tree and shrub layers) is fed on by many species, including	slugs
		snails
		millipedes
		springtails
		woodlice
		orobateid mites
		roundworms
		protozoa

FUNGI

are fed on by adults and larvae of many insects (for example, fungus beetle, boletus beetle)

The animals listed above all feed directly on plants or plant products; they are all classified as herbivores. Woodland, being rich in vegetation, supports a rich population of herbivores which, in turn, provides food for many carnivorous animals.

HERBIVORES

are fed on by insectivorous birds (blackbird, thrushes, green woodpecker)
insectivorous mammals (hedgehog, shrews)
insectivorous insects (robber flies, ladybirds)
spiders (wolf spider, cross spider)

Some of the insectivorous insects may themselves be eaten by birds. In turn, all the carnivores as well as some of the herbivores listed above may be eaten by the top carnivores, which include:

birds of prey (owls, kestrel)
beasts of prey (stoat, weasel, fox)

All the animals produce excreta, and this rich material provides food for animals such as:

dung beetles (dumble dor)
dung flies (yellow dung fly) and several other dipterous insects.

When the animal dies its body represents a valuable source of food, soon exploited by the carrion feeders such as:

larvae of blowfly
larvae of green-bottle
sexton beetles

There are probably no animals that do not harbour parasites. Birds and mammals carry in their fur or feathers an assortment of ectoparasites such as:

ticks
mites
fleas

Endoparasites live within the bodies of their hosts. There are so many of these that none deserves special mention. Unless a special study of parasitism is undertaken they are unlikely to be found, yet the existence of parasites must be recognized as an illustration of yet one more way in which the woodland provides food (and shelter) for its inmates.

The most characteristic way in which a wood provides shelter for animals is by offering comparatively safe nesting sites on the crowns of trees and an abundance of other places to rest or roost.

Squirrels and many birds take advantage of this protection. Some birds nest high in the crowns (rook, wood-pigeon, heron) while others habitually nest in the lower branches (wood warbler). Others nest among the branches of the smaller trees and bushes (great tits, song thrush, blackbird) and there are some that have preferences for the darker, denser bushes such as holly and yew (bullfinch). Holes in tree trunks provide security and protection from the weather for several bird species (blue-tit, robin, green woodpecker, tawny owl) and are sometimes used by squirrels. Fewer birds nest near the ground in hollow stumps, an exception being the coal-tit. The animals that feed inside the tree—the bark-borers and those that cause galls—also gain shelter by this habit. Many small insects, such as shield bugs, shelter by hiding in crevices in tree bark. When a tree has died or has been felled and is rotting, the moist soggy wood and bark is an ideal place for many animals. Some feed on the rotting wood, and others are afforded protection against desiccation by the damp conditions. Animals found on rotting stumps and logs include those primitive insects, the bristle-tails and silverfish; the terrestrial flatworm, *Rhynchodermis;* and woodlice.

The field-layer plants are not suitable as nesting sites, but they can act as a concealing cover for the nests of ground-nesting birds such as pheasant. Their main contribution is as hosts of gall insects, leaf miners and similar parasites. The action of insects that feed on leaves or suck the sap may cause the leaf to curl, and within the case so formed the insect is protected from attack by birds. In addition the plants of the field and ground layers create at ground level the still, humid atmosphere of moderate temperature so necessary to the survival of the many minute animals living in the leaf litter and on the soil surface.

Below this region, the soil itself provides equable conditions for a population of soil animals, including earthworms, centipedes, and woodlice, to mention only a few, but this part of the woodland pattern is best treated as a separate topic and will not be discussed further here.

FLORA OF DAMP OAKWOOD

(for abbreviations, see endpapers)

DESCRIPTION	SCIENTIFIC NAME	COMMON NAME	M[1]	D[2]	HABIT
1. Tree layer					
Dominant	*Quercus robur*	common oak, pedunculate oak	4-5	U	W
Occasionals	*Alnus glutinosa*	alder	2-3	U	W FBL
	Fagus sylvatica	beech	4-5	U	W
	Populus nigra	black poplar	4		W FBL
	Ulmus glabra	wych elm	2-3		W FBL
Margins/ clearings	*Acer pseudoplatanus*	sycamore	4-6	U	W
	Aesculus hippocastanum	horse chestnut	5-6	U	W
	Fraxinus excelsior	ash	4-5	U	W FBL
	Populus tremula	aspen	2-3		W FBL
	Tilia × europaea	common lime	7		W
2. Second tree layer					
	Ilex aquifolium	holly	5-8		W
	Malus sylvestris	crab-apple	5	S	W
	Prunus avium	gean, wild cherry	4-5		W
	Prunus padus	bird-cherry	5	N	W
	Taxus baccata	yew	3-4	+	W
3. Shrub layer					
	Carpinus betulus	hornbeam	4-5		W
	Corlyus avellana	hazel	1-4	U	W FBL
	Crataegus monogyna	hawthorn	5-6	U	W
	Euonymus europaeus	spindle-tree	5-6	S+	W
	Prunus spinosa	blackthorn, sloe	3-5	S	W
	Ribes uva-crispa	gooseberry	3-5		W
	Rosa canina	dog-rose	6-7		W
	Rubus fruticosus	blackberry, bramble	6-9	U	W
	Salix alba	white willow	4-5		W
	Salix caprea	goat willow, great sallow	3-4		W FBL
	Salix cinerea	common sallow	3-4		W FBL

[1] Months of flowering [2] Distribution

DESCRIPTION	SCIENTIFIC NAME	COMMON NAME	M	D	HABIT
	Salix fragilis	crack willow	4		W FWL
	Sambucus nigra	elder	6-7	U	W
	Thelycrania sanguinea	dogwood	6-7	S+	W
	Viburnum opulus	guelder rose	6-7		W
Climbers	*Clematis vitalba*	old man's beard, traveller's joy	7-8	S+	W
	Hedera helix	ivy	9-11	U	W
	Lonicera periclymenum	honeysuckle	6-9	U	W

4. Field layer

DESCRIPTION	SCIENTIFIC NAME	COMMON NAME	M	D	HABIT
Early spring flowers	*Anemone nemorosa*	wood anemone	3-5	U	R
	Caltha palustris	kingcup, marsh marigold	3-7	U	R
	Endymion non-scriptus	bluebell	4-6	U	B
	Euphorbia amygdaloides	wood-spurge	3-5	S	S
	Glechoma hederacea	ground ivy	3-5	U	S
	Helleborus foetidus	stinking hellebore	3-4		S
	Helleborus viridis	green hellebore	3-4		S
	Mercurialis perennis	dog's mercury	2-4	U	R
	Primula vulgaris	primrose	12-5	U	R
	Ranunculus ficaria	lesser celandine	3-5	U	T
	Veronica chamaedrys	germander speedwell	3-7	U	S
	Viola reichenbachiana	pale wood-violet	3-5	S	S
Spring flowers	*Adoxa moschatellina*	moschatel, townhall clock	4-5		R
	Allium ursinum	ramsons	4-6		B
	Anthriscus sylvestris	cow parsley, keck	4-6	U	S
	Arum maculatum	lords-and-ladies, cuckoo-pint	4-5	S	T
	Fragaria vesca	wild strawberry	4-7		S
	Luzula pilosa	hairy woodrush	4-6		S
	Orchis mascula	early purple orchid	4-6		T
	Oxalis acetosella	wood sorrel	4-5	U	R
	Ranunculus auricomus	goldilocks	4-5		S
	Stellaria holostea	greater stitchwort	4-6	U	S
	Viola riviniana	common violet	4-6	U	S

DESCRIPTION	SCIENTIFIC NAME	COMMON NAME	M	D	HABIT
Summer flowers or spores	*Ajuga reptans*	bugle	5-7	U	R
	Alchemilla vulgàris agg.	lady's mantle	6-8		S
	Angelica sylvestris	wild angelica	7-9	U	S
	Anthoxanthum odoratum	sweet vernal grass	4-6	U	S
	Arctium minus	lesser burdock	7-9	U	b
	Asplenium adiantum-nigrum	black spleenwort	6-10		R
	Athyrium filix-femina	lady fern	7-8		R
	Betonica officinalis	betony	6-9	S	R
	Blechnum spicant	hard fern	6-8		R
	Brachypodium sylvaticum	slender false-brome	7		S
	Carex flacca	carnation-grass	5-6		R
	Carex pendula	pendulous sedge	5-6	S	R
	Carex sylvatica	wood sedge	5-7	R	R
	Chamaenerion angustifolium	rose-bay willow-herb	7-9	U	S
	Chrysosplenium oppositifolium	opposite-leaved golden saxifrage	5-7		S
	Circaea lutetiana	enchanter's night-shade	6-8		R
	Cirsium palustre	marsh thistle	7-9	U	b
	Conopodium majus	pignut, earthnut	5-6		T
	Dactylorchis fuchsii	common spotted orchid	6-8		T
	Deschampsia caespitosa	tufted hair-grass	6-8	U	S
	Dryopteris dilatata	broad buckler-fern	7-9	U	R
	Dryopteris filix-mas	male fern	7-9	U	R
	Epilobium adnatum	square-stemmed willow-herb	7-8	S	S
	Epilobium montanum	broad-leaved willow-herb	6-8	U	S
	Epilobium obscurum	dull-leaved willow-herb	7-8		S
	Festuca gigantea	tall brome	6-7		S
	Filipendula ulmaria	meadow-sweet	6-9	U	R
	Galeobdolon luteum	yellow archangel	5-6	S	S
	Galium mollugo	great hedge bedstraw	6-7		S
	Galium odoratum	sweet woodruff	5-6		R

DESCRIPTION	SCIENTIFIC NAME	COMMON NAME	M	D	HABIT
	Geranium robertianum	herb Robert	5-9	U	a
	Geum urbanum	wood avens	6-8	U	R
	Heracleum sphondylium	cow parsnip, hogweed, keck	7-8	U	S
	Holcus lanatus	Yorkshire fog	6-9	U	S
	Hypericum perforatum	common St John's wort	6-9	S	R
	Iris pseudacorus	yellow flag	5-7	U	R
	Juncus conglomeratus	conglomerate rush	5-7		R
	Juncus effusus	soft rush	6-8	U	R
	Lapsana communis	nipplewort	7-9	U	a
	Listera ovata	twayblade	6-7		R
	Lolium perenne	rye-grass	5-8	U	R
	Lychnis flos-cuculi	ragged Robin	5-6	U	S
	Lysimachia nemorum	yellow pimpernel	5-9		S
	Melica uniflora	wood melick	5-6		S
	Myosotis scorpioides	water forget-me-not	5-9	U	S
	Myosotis sylvatica	wood forget-me-not	5-6	U	S
	Phyllitis scolopendrium	hart's-tongue fern	7-8		R
	Polypodium vulgare agg.	polypody	6-9		R
	Potentilla erecta	common tormentil	6-9	U–	S
	Potentilla sterilis	barren strawberry	2-5		S
	Prunella vulgaris	self-heal	7-9		R
	Pteridium aquilinum	bracken	7-8	U	R
	Ranunculus repens	creeping buttercup	5-8	U	S
	Rumex acetosella	sorrel	5-6	U	S
	Sagina procumbens	procumbent pearlwort	5-9	U	S
	Sanicula europaea	sanicle	5-9	+	S
	Scrophularia aquatica	water betony	6-9	S	R
	Scrophularia nodosa	figwort	6-9		R
	Silene dioica	red campion	5-6	U	b/S
	Solanum dulcamara	bittersweet, woody nightshade	6-9	S	W
	Stachys sylvatica	hedge woundwort	7-8	U	R
	Stellaria graminea	lesser stitchwort	5-8		S
	Succisa pratensis	Devil's-bit scabious	6-10	U	S
	Tamus communis	black bryony	5-7	S	T
	Teucrium scorodonia	wood sage	7-9		R

DESCRIPTION	SCIENTIFIC NAME	COMMON NAME	M	D	HABIT
	Torilis japonica	upright hedge-parsley	7-8	U	a
	Urtica dioica	stinging nettle	6-8	U	Ro
	Veronica officinalis	common speedwell	5-8		S
	Zerna ramosa	hairy brome	7-8		S

5. Ground layer

DESCRIPTION	SCIENTIFIC NAME	COMMON NAME	M	D	HABIT
Liverworts	*Diplophyllum albicans*				LLW
	Lepidozia reptans				LLW
	Lophocolea bidentata				LLW
	Pellia epiphylla	wide-nerved liverwort		U	TLW
	Plagiochila asplenoides	spleenwort scale-moss			LLW
	Scapania nemorosa				LLW
Mosses	*Atrichum undulatum*	wavy-leaved thread-moss		U	CM
	Aulacomnium androgynum	bud-headed thread-moss			CM
	Dicranella heteromalla	silky fork-moss		U	CM
	Dicranum scoparium	lesser fork-moss			CM
	Eurhynchium praelongum	long trailing fork-moss			FM
	Fissidens bryoides	common flat fork-moss			CM
	Mnium affine				CM
	Mnium hornum	swan's-neck thread-moss		U	CM
	Mnium punctatum	dotted thread-moss			CM
	Mnium undulatum	palm tree-moss			CM
	Plagiothecium denticulatum	sharp fern-like feather-moss			FM
	Plagiothecium undulatum	waved feather-moss			FM
	Pseudoscleropodium purum	neat meadow feather-moss			FM
	Rhytidiadelphus squarrosus	drooping-leaved feather-moss			FM
	Thamnium alopecurum	fox-tail feather-moss			FM

DESCRIPTION	SCIENTIFIC NAME	COMMON NAME	M	D	HABIT
	Thuidium tamarascinum	tamarisk-leaved feather-moss			FM
Algae		various			
Ferns		prothalli			

6. Epiphytes on the bark of trees

DESCRIPTION	SCIENTIFIC NAME	COMMON NAME	M	D	HABIT
Ferns	*Polypodium vulgare* agg.	polypody	6-9		R
Mosses	*Aulacomnium andrigynum*	bud-headed thread-moss			CM
	Camptothecium sericeum	silky wall feather-moss			FM
	Dicranoweissia cirrata	curly thatch-moss			CM
	Eurhynchium praelongum	long trailing fork-moss			FM
	Hypnum cupressiforme	cypress-leaved feather-moss		U	FM
	Isopterygium elegans	elegant feather-moss			FM
	Neckera complanata	flat feather-moss			FM
	Orthotrichum affine				CM
	Orthotrichum diaphanum				CM
	Orthotrichum lyelli				CM
	Tetraphis pellucida				CM
Liverworts	*Frullania dilatata*				LLW
	Lophocolea cuspidata				LLW
	Lophocolea heterophylla				LLW
Lichens	*Alectoria fuscescens*			N	BL
	Buellia canescens				CL
	Cetraria glauca				LL
	Cladonia coniocraea			U	LL
	Evernia prunastri	mousse de chêne			BL
	Graphis elegans			S	CL
	Hypogymnia physodes			U	LL
	Lecanora conizaeoides				CL
	Lecanora expallens			U	CL
	Lecidea limitata			U	CL

DESCRIPTION	SCIENTIFIC NAME	COMMON NAME	M	D	HABIT
	Lepraria incana				CL
	Parmelia laevigata			S	LL
	Parmelia saxatilis	crottle		U	LL
	Parmelia subrudecta			S	LL
	Parmelia sulcata			U	LL
	Pertusaria amara			U	CL
	Pertusaria pertusa			U	CL
	Physcia leptalea				LL
	Ramalina farinacea				BL
	Usnea ceratina			S	BL
	Usnea rubiginea			U	BL
	Usnea subfloridans				BL
	Xanthoria parietina			U	CL
Algae	*Hormidium nitens*				
	Pleurococcus vulgaris			U	

7. Fungi

	Collybia dryophila				CF
	Fistulina hepatica	beefsteak fungus			BF
	Grifolia gigantea				BF
	Grifolia sulphurea				BF
	Lactarius quietus				CF
	Mycaena inclinata				CF
	Mycaena polyadelpha				CF
	Polyporus squamosus	dryad's saddle			BF
	Russula sorsoria				CF
	Russula vesca				CF
	Tricholoma sulphureum				CF

Dry oakwood

The characteristic dominant tree is *Quercus petraea*, the durmast oak or sessile oak, which is distinguished from *Quercus robur* of the damp oakwoods by several features, especially the fact that the acorn cup is sessile, being attached directly to the twig. Dry oakwoods are found on the lighter, drier soils which are formed from sandstone or slate and some of the older rocks of Britain. This type of soil occurs mainly in the west and north of Britain, including the Pennines, with additional areas in south-east England on the gravels and sands of Hertfordshire and the sand of Surrey and Kent. Thus, the plants characteristic of this kind of woodland show a distinctive distribution which might be described as 'north and west plus south-east'. Of those listed on pp. 27-28, the following show this distribution:

Quercus petraea
Calluna vulgaris
Erica cinerea
Vaccinium myrtillus
Blechnum spicant
Deschampsia flexuosa
*Dryopteris borreri**
Galium saxatile
Hypericum pulchrum
Luzula sylvatica
Lysimachia nemorum
Solidago virgaurea

It will be noticed that relatively few plants with 'universal' distribution are listed, and none with northerly or southerly distributions. This emphasizes the fact that the occurrence of this type of woodland is dependent on soil conditions, rather than on climatic or other conditions.

Damp oakwood and dry oakwood represent two contrasting types, but many of the woods found fall into neither category. In northern districts, *Q. robur* may be seen growing among *Q. petraea* in the valleys, though not higher up the slopes. Hybrids between the two may sometimes be found. The heathy oakwoods are a third type, similar in some ways to the dry oakwood, and growing on heathy soils that are poor and acid, and usually sandy. In the north, these woods are dominated by *Q. petraea*, but in the south (Kent, Hampshire, London basin) they are dominated by either oak species, or by both. For these reasons it is often hard to decide into which category a particular oakwood should fall. Fortunately, this is of no great concern here, for our aim is not to go into finer details, but to extract those ecological and biological ideas that stand out most clearly. A certain amount of simplification is necessary, so that apart from being ready to recognize that intermediates and variations do exist we will consider the dry and damp oakwoods as being two distinct types depending on distinct sets of soil conditions.

Since the rocks from which dry oakwood soils originate are primarily siliceous, the soils produced consist of a large amount of chemically unreactive silica. These contain little mineral material, and what little they have is soon leached away by drainage. These soils drain easily, unless underlain

* One species of the *Dryopteris filix-mas* aggregate.

by clay or shale, which is seldom since they are mostly in hilly areas. There is a rapid run-through of rainwater, carrying the minerals below, out of reach of the roots of the plants. Since these soils are mainly in the west of Britain, where the highest rainfall occurs, the leaching effect is all the more apparent. The soils are formed mainly on hillsides and therefore tend to be shallow. Contrast these with the deep, rich loams and clays on which damp oakwood is found. One mineral of special interest is calcium. Like other minerals this is leached from the soil, so these soils, either because they were low in calcium to begin with or have lost the calcium they once had, become suitable for the calcifuge plants, those plants that thrive best on calcium-deficient soils. The dry woodland flora contains a high proportion of such plants (p. 27). *Pteridium aquilinum*, which is a widespread plant found also in damp oakwood in small amounts, becomes the dominant field-layer plant on the calcium-free soils of many dry oakwoods.

The pattern of plant and animal life in dry oakwood is much the same as that already described for damp oakwood, and these points will not be repeated here. The main differences arise from the nature of the soil, which is dry and shallow, so that the oaks and other trees cannot root deeply. Perhaps partly for this reason the other plant layers are poorly developed, for the roots of the trees compete directly with the roots of the field-layer plants for the scanty supply of water and minerals. The tree layer itself may be sparse, especially in hilly districts where the thin soil will restrict the growth of trees and the oaks may have the stature of bushes. There is generally no second tree layer, and the shrub layer is poorly developed. *Calluna* or *Ilex* are often dominant among the shrubs. Note that *Ilex* is included among the shrubs of this list, whereas in the damp oakwood it was listed in the second tree layer. This reflects the reduction in its stature owing to poor soil conditions. Similarly, *Sorbus aucuparia*, which can grow to a height of 20 m, is a shrubby plant in dry oakwood. Four of the shrub-layer plants are low, spreading bushy plants. These four are *Calluna*, *Erica*, *Sarothamnus*, and *Vaccinium*. Like *Ilex*, they all show xeromorphic characters, which fit them for life in dry soils. The relative amounts of these shrubs present in any given wood will depend on several conditions. In general, the xeromorphs are commoner in higher and more exposed woods. *Vaccinium* is more tolerant of exposure than *Calluna* and may largely replace it under exposed conditions. It may also be found in woods that are too shady to allow *Calluna* to grow.

Beneath the shrub layer there is a field layer, but this lacks the variety and luxuriance of the field layer of the damp oakwood. The profusion of spring flowers is absent. The only spring-flowering plant is *Endymion non-scriptus* (bluebell, not to be confused with *Campanula rotundifolia* which in Scotland is also called bluebell). This may be abundant in some woods. Later the fronds of *Pteridium* grow up, obscuring the dying leaves of the bluebell, and *Pteridium* becomes the dominant member of the field layer. This is unusual among ferns in that its fronds die back in autumn, covering the woodland with a thick carpet of brown leaves. Before the next season's leaves grow up the bluebells are able to leaf and flower again.

As in the damp oakwood, certain species may be better adapted for survival under special conditions. Several require greater light intensity, so are to be found only at the wood margins or in open spaces. Among these are:

Centaurium erythraea *Potentilla erecta*
Digitalis purpurea

This list includes the only annual and biennial plants from the complete list.

Other species may be better adapted for life in dry-oakwood conditions in ways which are less easy to discover. By comparing lists it will be seen that several *genera* are present in both lists, but are represented by different *species*:

DAMP OAKWOOD	DRY OAKWOOD
Quercus robur	*Quercus petraea*
Deschampsia caespitosa	*Deschampsia flexuosa*
Galium mollugo *Galium odoratum*	*Galium saxatile*
Holcus lanatus	*Holcus mollis*
Hypericum perforatum	*Hypericum pulchrum*
Ranunculus repens	*Ranunculus bulbosus*

However, these lists indicate only a *tendency* for one species to be replaced by the other. This is by no means clear-cut, and both species of one genus may be found growing in the same wood. Some woods, as already mentioned, have both species of *Quercus*, and show vegetation intermediate between the damp oakwood and dry oakwood types.

Ground-layer plants differ little between the two types of oakwood, or perhaps the differences have not been investigated so exactly. Certain mosses such as *Zygodon viridissimus* (green tufted thread-moss) and *Hylocomium splendens* (glittering feather-moss) are commoner on soils deficient in calcium and so may be found more commonly in dry oakwoods. Most species seem less affected by the soil conditions, and presumably the shallow depth of dry oakwood soils has little influence on ground flora which send their rhizoids into only the top few millimetres of soil. They compete with the roots of the field-layer plants no more than they would in a damp oakwood. Thus, many of the species listed on pp. 21-22 are found in dry oakwoods. The epiphytes too are not directly affected by soil conditions, so the species found will be similar. Possibly because the dry oakwoods are located mainly in the west, which has heavier rainfall, and possibly because areas such as south-west England and Scotland are far from large centres of population and air pollution, woods in these areas show abundant growth of lichens.

FLORA OF DRY OAKWOOD

(for abbreviations, see endpapers)

DESCRIPTION	SCIENTIFIC NAME	COMMON NAME	M[1]	D[2]	HABIT
1. Tree layer					
Dominant	*Quercus petraea*	durmast oak, sessile oak	4-5		W
Other trees	*Betula pubescens*	birch	4-5		W
	Betula pendula	silver birch	4-5	U	W
	Quercus robur	common oak, pedunculate oak	4-5	U	W
	Ulmus glabra	wych elm	2-3		W FBL
2. Second tree layer—usually absent					
3. Shrub layer					
	Calluna vulgaris	ling, heather	7-9	—	W
	Carpinus betulus	hornbeam	4-5		W
	Crataegus monogyna	hawthorn	5-6	U	W
	Erica cinerea	bell-heather	7-9	—	W
	Ilex aquifolium	holly	5-8		W
	Rosa canina	dog rose	6-7		W
	Rubus fruticosus	blackberry, bramble	6-9	U	W
	Sarothamnus scoparius	broom	5-6		W
	Sorbus aucuparia	mountain ash	5-6		W
	Vaccinium myrtillus	bilberry, blaeberry, whortleberry, huckleberry	4-6	—	W
4. Field layer					
Dominant	*Pteridium aquilinum*	bracken	7-8	U—	R
Other plants	*Asplenium adiantum-nigrum*	black spleenwort	6-10		R
	Blechnum spicant	hard fern	6-8		R
	Campanula rotundifolia	harebell, bluebell	7-9	U	S
	Centaurium erythraea	common centaury	6-10		a
	Deschampsia flexuosa	wavy hair-grass	6-7	—	S
	Digitalis purpurea	foxglove	6-9	U	b
	Dryopteris borreri	male fern	7-8		R

[1] Months of flowering [2] Distribution

DESCRIPTION	SCIENTIFIC NAME	COMMON NAME	M	D	HABIT
	Endymion non-scriptus	bluebell	4-6	U	B
	Galium saxatile	heath bedstraw	6-8		S
	Holcus mollis	creeping softgrass	6-7		S
	Hypericum pulchrum	slender St John's wort	6-8		R
	Luzula sylvatica	greater woodrush	5-6		S
	Lysimachia nemorum	yellow pimpernel	5-9		S
	Mercurialis perennis	dog's mercury	2-4	U	R
	Moehringia trinervia	three-nerved-sand-wort	5-6		a
	Oxalis acetosella	wood sorrel	4-5		R
	Phyllitis scolopendrium	hart's-tongue fern	7-8		R
	Potentilla erecta	creeping tormentil	6-9	U–	S
	Primula vulgaris	primrose	12-5		S
	Ranunculus bulbosus	bulbous buttercup	5-6	U	C
	Ranunculus ficaria	lesser celandine	3-5	U	T
	Solidago virgaurea	golden rod	7-9		S
	Teucrium scorodonia	wood sage	7-9		R
	Veronica officinalis	common speedwell	5-8		S

plus occasional plants from the list of pp. 18-21.

5. Ground layer

DESCRIPTION	SCIENTIFIC NAME	COMMON NAME	M	D	HABIT
Liverworts	*Calypogeia fissa*				LLW
	Diplophyllum albicans				LLW
	Lophocolea bidentata				LLW
	Lophocolea heterophylla				LLW
	Pellia epiphylla	wide-nerved liver-wort			TLW
	Plagiochila asplenoides	spleenwort scale-moss			LLW
Mosses	*Dicranella heteromalla*	silky fork-moss		U	CM
	Eurhynchium praelongum	long trailing fork-moss			FM
	Leucobryum glaucum	white fork-moss			CM
	Mnium punctatum	dotted thread-moss			CM
	Thamnium alopecurum	fox-tail feather-moss			FM
Algae		various			
Ferns		prothalli			

6. Epiphytes—see list on pp. 22-23.

7. Fungi—see list on p. 23.

Beechwood

The beech can be planted and grown successfully in almost any area in Britain, except on heavy, wet clays, but beechwoods are mainly found on the dry chalky soils of south-east England and on the limestone soils of the Cotswolds. As with oakwoods, it is possible to describe several distinct types, depending on the nature of the soil, its depth, and its slope. Here discussion will be confined to those features that are characteristic of beechwoods in general. The flora lists (p. 31) include species found in one or more of the types of beechwood, so that *all* of these would not necessarily be found in a *single* wood.

The most notable feature of a beechwood is the heavy shade cast by the leaves of the beech. The leaves are arranged on the twig in mosaic patterns so that each leaf fits as far as possible into the spaces between the leaves above it. Consequently, little light can penetrate the crown of a beech tree. Further, the leaves of beech open about two weeks earlier than those of oak, so that the growing season for plants in the beechwood is shortened, and at a time when a reduction of two weeks makes a great deal of difference. The result is that, below the canopy of the beechwood, growth of the lower layers of plants is sparse or non-existent, except, of course, around the wood margins and in open spaces. A second tree layer is usually not found, and shrubs, if any, are few and small. In effect, the leaves of the lower branches of the beech tree take the place of the second tree layer, for they can tolerate dense shade themselves. *Rubus fruticosus* is able to grow in shade, and this sometimes becomes the dominant member of the shrub layer in woods growing on deep, loamy soils. Functionally, *Rubus* is a member of the field layer, so one can even then consider the shrub layer to be absent. In the shadier parts of such woods *Rubus* is replaced by *Oxalis acetosella* as dominant field-layer plant.

The field layer is best developed in early spring, when the spring-flowering plants are active before the leaf canopy closes. Later in the year, when their leaves have died down, large areas of soil may be completely bare, the more so in closely planted woods with the heaviest cover. Another factor contributing to the sparsity of field-layer plants is the nature of the root system of the beech. Whereas the oaks are deep-rooting, the beech forms a dense branching system of rootlets throughout the surface layers of soil, and these remove large quantities of water which would otherwise be available for field-layer plants.

In beechwoods the soil is usually covered with a layer of dead beech leaves several centimetres thick. Depending on local conditions this layer consists of dry, loose leaves on the surface with a rich humus layer beneath, or of a densely packed layer of sodden leaves, again with humus below. Where the leaves are loose, it is impossible for mosses and other members of the ground flora to become established, so the ground layer is absent too. Even bare patches of soil are not suitable for colonization, for a change of wind direction may soon cover such an area with a drift of leaves. The bark of beech is thin, and does not encourage the growth of mosses, liverworts, lichens or algae, except in regions of the trunk which are most thoroughly wetted as rain-water drains down. The few mosses and liverworts that become established are usually found at the bases of tree trunks and on the bark of the larger tree roots, which spread out above the soil surface.

Conditions of low light intensity with ample supplies of humus favour saprophytic organisms, and two types of saprophyte are common in beechwoods. Most of the flowering plants rely on chlorophyll for manufacturing organic materials by photosynthesis, but a few are adapted to the saprophytic mode of life. In beechwoods, two of these may be found growing in the humus layer—*Neottia nidus-avis*, with its brownish scale-like leaves and flowers, and *Monotropa hypopitys*, with its yellowish-white leaves and flowers. The more commonly-met saprophytes, the fungi, are frequently found in beechwoods, especially in late summer and autumn, where they grow on the thick humus layer. Some typical species are listed on pp. 32-33. The list includes *Amanita phalloides*, the death cap, which is common in beechwoods. Since it is fatally poisonous, it is as well to be able to recognize this fungus, with its pale olive-green cap and its white stalk and gills, and to avoid eating it or any other fungus that fits this description. *Ganoderma applanatum* is a bracket fungus commonly found on beech trees.

FLORA OF BEECHWOOD

(for abbreviations, see endpapers)

DESCRIPTION	SCIENTIFIC NAME	COMMON NAME	M[1]	D[2]	HABIT
1. Tree layer					
Dominant	*Fagus sylvatica*	beech	4-5	U	W
Other trees	*Acer pseudoplatanus*	sycamore	4-6	U	W
	Fraxinus excelsior	ash	4-5	U	W FBL
	Ilex aquifolium	holly	5-8		W
	Prunus avium	gean, wild cherry	4-5		W
	Quercus petraea	durmast oak, sessile oak	4-5	U	W
	Quercus robur	common oak, pedunculate oak	4-5	U	W
	Salix aurita	eared sallow	4		W
	Sorbus aria	whitebeam	5-6	S	W
	Taxus baccata	yew	3-4	+	W
2. Shrub layer (if present)					
	Acer campestre	field maple	5-6	S	W
	Calluna vulgaris	ling, heather	7·9	—	W
	Corylus avellana	hazel	1-4	U	W FBL
	Euonymus europaeus	spindle-tree	5-6	S+	W
	Hedera helix	ivy	9-11	U	W
	Rubus fruticosus	blackberry, bramble	6-9	U	W
	Ruscus aculeatus	butcher's broom	1-4	S	W
	Sambucus nigra	elder	6-7	U	W
3. Field layer (if present)					
	Ajuga reptans	bugle	5-7	U	R
	Anemone nemorosa	wood anemone	3-5	U	R
	Arum maculatum	lords-and-ladies, cuckoo-pint	4-5	S	T
	Blechnum spicant	hard fern	6-8		R
	Brachypodium sylvaticum	slender false-brome	7		S
	Carex sylvatica	wood sedge	5-7		R
	Cephalanthera damasonium	white helleborine	5-6	S+	R

1 Months of flowering 2 Distribution

DESCRIPTION	SCIENTIFIC NAME	COMMON NAME	M	D	HABIT
	Circaea lutetiana	enchanter's night-shade	6-8		R
	Deschampsia caespitosa	tufted hair-grass	6-8	U	S
	Deschampsia flexuosa	wavy hair-grass	6-7	–	S
	Epipactis helleborine	broad helleborine	7-10	S	R
	Euphorbia amygdaloides	wood spurge	3-5	S	S
	Fragaria vesca	wild strawberry	4-7		S
	Galeobdolon luteum	yellow archangel	5-6		S
	Galium odoratum	sweet woodruff	5-6		R
	Geranium robertianum	herb Robert	5-9	U	a
	Hypericum pulchrum	slender St John's wort	6-8		R
	Mercuralis perennis	dog's mercury	2-4		R
	Monotropa hypopitys	yellow bird's nest	6-8		Sap
	Neottia nidus-avis	bird's nest orchid	6-7		Sap
	Oxalis acetosella	wood sorrel	4-5	U	R
	Sanicula europaea	Sanicle	5-9		S
	Veronica chamaedrys	germander speedwell	3-7	U	S
	Zerna ramosa	hairy brome	7-8		S

4. Ground layer/Epiphytes

DESCRIPTION	SCIENTIFIC NAME	COMMON NAME	M	D	HABIT
Liverworts	*Frullania dilatata*				LLW
	Lophocolea bidentata				LLW
	Madotheca platyphylla				TLW
	Metzgeria furcata	flat-leaved scale-moss			LLW
Mosses	*Camptothecium sericeum*	silky wall feather-moss			FM
	Eurhynchium striatum	fork-moss			FM
	Hypnum cupressiforme	cypress-leaved feather-moss			FM
	Leucobryum glaucum	white fork-moss			CM
	Mnium undulatum	dotted thread-moss			CM

5. Fungi

DESCRIPTION	SCIENTIFIC NAME	COMMON NAME	M	D	HABIT
	Amanita citrina				CF
	Amanita phalloides	death cap			CF

DESCRIPTION	SCIENTIFIC NAME	COMMON NAME	M	D	HABIT
	Ganoderma applanatum				BF
	Inocybe fastigiata				CF
	Mycena spp.				CF
	Russula spp.				CF

Ashwood

Ash has been commonly planted in hedgerows and in mixed woodland, but woods dominated mainly by ash are less widespread. Ash grows best on calcareous soils, and it is in these regions that we find ashwoods. In chalky and limestone areas in central and southern Britain the ashwoods are one stage in the succession that lead to a beechwood. The beech trees are slower growing, but in southern Britain they eventually over-top the ash trees and then become the dominant tree. Further north, especially in the Derbyshire dales and in West Yorkshire, beech does not grow as well, and in these regions permanent ashwoods develop.

The canopy of ash trees is relatively open, especially when compared with that of beech, so that beneath these trees a wide variety of plants can grow. Also, since ash when it reaches full stature (about 25 m) is not one of the largest among our common native trees, no distinct second tree layer is formed. Trees of other species, particularly *Ulmus glabra* may frequently be found growing to the same height as the ash. Beneath the open tree layer a variety of shrubs can flourish, including several that prefer calcareous soils (calcicoles, indicated by + on the lists). The same remarks apply to the field-layer plants, which grow profusely and in great variety. As in all woodlands, the field-layer plants are almost exclusively perennials. Mosses are not so abundant here as in oakwoods, presumably because the bark of the ash is less suited to their growth. On the ground several species may be found, including some that are more or less calcicolous.

FLORA OF ASHWOOD

(for abbreviations, see endpapers)

DESCRIPTION	SCIENTIFIC NAME	COMMON NAME	M[1]	D[2]	HABIT
1. Tree layer					
Dominant	*Fraxinus excelsior*	ash	4-5	U	W FBL
Other trees	*Acer campestre*	field maple	5-6	S	W
	Euonymus europaeus	spindle-tree	5-6	S+	W
	Fagus sylvatica	beech	4-5	U	W
	Populus tremula	aspen	2-3		W
	Prunus padus	bird-cherry	5	N	W
	Quercus robur	common oak, pedunculate oak	4-5	U	W
	Sorbus aria	white beam	5-6	S	W
	Ulmus glabra	wych elm	2-3		W FBL
2. Shrub layer					
	Clematis vitalba	traveller's joy, old man's beard	7-8	S+	W
	Corylus avellana	hazel	1-4	U	W FBL
	Crataegus monogyna	hawthorn	5-6	U	W
	Hedera helix	ivy	9-11	U	W
	Ligustrum vulgare	privet	6-7		W
	Rhamnus catharticus	buckthorn	5-6	S+	W
	Rosa canina	dog rose	6-7		W
	Rubus caesius	dewberry	6-9	S	W
	Rubus saxatilis	stone bramble	6-8	N	W
	Salix cinerea	common sallow	3-4		W FBL
	Sambucus nigra	elder	6-7	U	W
	Taxus baccata	yew	3-4	+	W
	Thelycrania sanguinea	dogwood	6-7	S+	W
	Viburnum lantana	wayfaring tree, mealy guelder rose	5-6	S+	W

[1] Months of flowering [2] Distribution

DESCRIPTION	SCIENTIFIC NAME	COMMON NAME	M	D	HABIT

3. Field layer

DESCRIPTION	SCIENTIFIC NAME	COMMON NAME	M	D	HABIT
	Adoxa moschatellina	moschatel, townhall clock	4-5		R
	Allium ursinum	ramsons	4-6		B
	Anemone nemorosa	wood anemone	3-5	U	R
	Arum maculatum	lords-and-ladies, cuckoo-pint	4-5	S	T
	Asplenium adiantum-nigrum	black spleenwort	6-10		R
	Campanula latifolia	large campanula, great bellflower	7-8		S
	Cirsium heterophyllum	melancholy thistle	7-8	N	S
	Convallaria majalis	lily-of-the-valley	5-6	+	R
	Dactylorchis maculata agg.	spotted orchid	6-8		T
	Deschampsia caespitosa	tufted hair-grass	6-8	U	S
	Dryopteris filix-mas agg.	male fern	7-8	U	R
	Euphorbia amygdaloides	wood spurge	3-5	S	S
	Fragaria vesca	wild strawberry	4-7		S
	Galeobdolon luteum	yellow archangel	5-6		S
	Galium odoratum	sweet woodruff	5-6		R
	Geum urbanum	wood avens	6-8	U	R
	Glechoma hederacea	ground ivy	3-5	U	S
	Hypericum hirsutum	hairy St John's wort	7-8	+	S
	Mercurialis perennis	dog's mercury	2-4	U	R
	Myosotis sylvatica	wood forget-me-not	5-6	U	S
	Oxalis acetosella	wood sorrel	4-5	U	R
	Polypodium vulgare agg.	polypody	6-9		R
	Ranunculus ficaria	lesser celandine	3-5	U	T
	Teucrium scorodonia	wood sage	7-9		R
	Valeriana officinalis	valerian	6-8	U	R

DESCRIPTION	SCIENTIFIC NAME	COMMON NAME	M	D	HABIT

4. Ground layer/Epiphytes

DESCRIPTION	SCIENTIFIC NAME	COMMON NAME	M	D	HABIT
Liverworts	*Frullania dilatata*				LLW
	Lopocolea heterophylla				LLW
	Madotheca platyphylla	flat-leaved scale-moss		+	LLW
	Plagiochila asplenoides	spleenwort scale-moss			LLW
Mosses	*Atrichum undulatum*	Catherine's moss, wavy-leaved thread-moss			CM
	Ctenidium molluscum	plumy-crested feather-moss		+	FM
	Eurhynchium striatum			+	FM
	Fissidens bryoides	common flat fork-moss			CM
	Hylocomium splendens	glittering feather-moss		+	FM
	Mnium longirostrum	long-beaked thread-moss		+	CM
	Plagiothecium denticulatum	sharp fern-like feather-moss			FM
	Thuidium tamariscinum	tamarisk-leaved feather-moss			FM
Lichens	many different species, especially of *Lecanora*, *Lecidea*, *Parmelia*, *Ramalina*, *Usnea* and *Xanthoria*				
Algae	*Pleurococcus vulgaris*			U	

Birchwood

This is found particularly in heathland areas, such as the heaths and commons of Surrey, and on moorland in Scotland. In general, birchwood is not a stable vegetation pattern but a stage in colonization and the production of woodland that might eventually become dominated by oak or beech. In the early stages of colonization, birch becomes established because it produces seed abundantly, because its seeds are light and easily transported to the new areas of colonization, and because it can tolerate poor soil. Later, the heavier seeds of oak or beech may arrive, and these trees, though growing slowly at first, will in the end overgrow the birch colonists. The birch species most often found is *Betula pubescens*, which is very tolerant of wet and cold; *Betula pendula* is more often found on drier soils. Beneath the tree layer and between the individual trees (for this type of wood is very open) a varied shrub layer and field layer can grow. The species making up this layer depend very much on what type of region is in the process of colonization and particularly the kind of soil. This is why the list of field-layer plants has been given in two sections (pp. 39-40). Note the relatively larger number of grass plants in these lists. A birchwood is an area in the process of being covered by tree canopy, and at the beginning one would expect to find growing there plants that are not typically found in mature woods. The grasses and the heath plants are examples. In fact, in these woods one may commonly find species that are also found growing in non-wooded areas in the same district.

The bark of birch is thin, acid in nature, and peels away frequently, so that it is unsuitable for epiphytes. For this reason lichens and algae are scarce and mosses few. In boggy or other areas of moist soil certain mosses may be plentiful on the soil but not on the trees. A number of fungi are associated with birch woods, including those listed on p. 41.

Plate 1. Oakwood, Derbyshire, April. Spring growth has not yet started. This is *Quercus robur* woodland but, being on shallow sloping soil with outcrops of rock (A), it is not a typical damp oakwood. It illustrates one of the many ways in which the basic pattern may be modified by local environmental conditions. The trees on this relatively poor soil are widely spaced and small, and the shrub layer is very sparse. In such open woodland grasses can grow, and their dominance of the field layer is assured by the grazing of sheep (B), which are able to enter the unfenced wood from the surrounding moorland. In small scattered areas of the wood (and throughout fenced woods in the neighbourhood) typical shrub-layer plants, such as bramble and bracken, occur. Even then their growth is sparse, presumably because of competition with the roots of trees in the shallow soil layer. The odd patches of shrub vegetation (C) in this wood are mainly on the most steeply sloping ground. Perhaps this is because sheep are unable to graze in these less accessible areas—this is a hypothesis that needs testing by field experiment. In the foreground there is a long hollow in the ground where water collects and the soil is moister. Under these conditions the dominant plants are sedges (D).

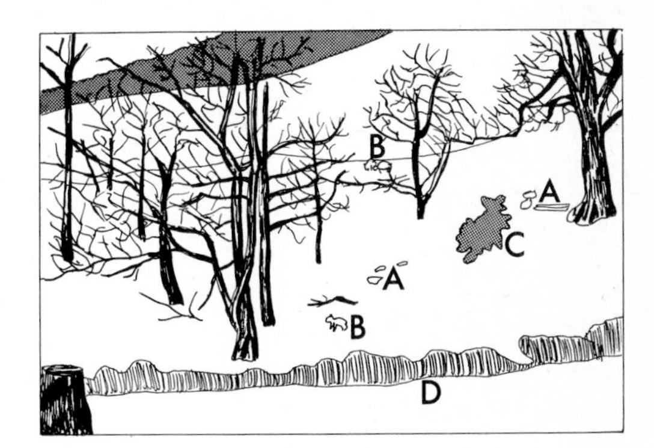

Plate 2. Pine plantation, Nottinghamshire, April. This is in a fertile river valley. The trees are closely planted and their crowns intermingle, casting heavy shade—in marked contrast to the wood shown in Plate 1. Another important point of contrast is that pines are evergreen, so they cast shade all the year round. This photograph was taken within a few days of that shown in Plate 1, yet compare the intensity of illumination within the two woods. The low light intensity prevents the growth of a shrub layer and field layer. The meadow vegetation seen in the foreground can extend into the wood only as far as the slanting rays of the sun early and late in the day can reach (broken line). In this marginal region a few shrubs (for example, the sycamore saplings, A) can grow.

FLORA OF BIRCHWOOD

(for abbreviations, see endpapers)

DESCRIPTION	SCIENTIFIC NAME	COMMON NAME	M[1]	D[2]	HABIT
1. Tree layer					
Dominant	*Betula pubescens*	birch	4-5		W
	or *Betula pendula*	silver birch	4-5		W
Common	*Sorbus aucuparia*	mountain ash	5-6		W
2. Shrub layer					
	Corylus avellana	hazel	1-4	U	W FBL
	Hedera helix	ivy	9-11	U	W
	Lonicera periclymenum	honeysuckle	6-9	U	W
	Populus tremula	aspen	2-3		W
	Prunus padus	bird-cherry	5	N	W
	Salix caprea	great sallow, goat willow	3-4		W
3. Field layer					
Lowland birchwood —good soils	*Agrostis tenuis*	common bent-grass	6-8	U	R
	Brachypodium sylvaticum	slender false-brome	7		S
	Dactylis glomerata	cock's-foot	5-7		S
	Endymion non-scriptus	bluebell	4-6	U	B
	Festuca rubra	red fescue	5-7	U	S
	Fragaria vesca	wild strawberry	4-7		S
	Holcus mollis	creeping soft-grass	6-7		S
	Luzula pilosa	hairy woodrush	4-6		S
	Lysimachia nemorum	yellow pimpernell	5-9		S
	Oxalis acetosella	wood sorrel	4-5	U	R
	Poa trivialis	rough-stalked meadow-grass	6		S
	Primula vulgaris	primrose	12-5		R
	Prunella vulgaris	self-heal	7-9		R
	Silene dioica	red campion	5-6	U	b/S
	Stellaria holostea	greater stitchwort	4-6	U	S
	Veronica chamaedrys	germander speedwell	3-7	U	S
	Viola riviniana	common violet	4-6	U	S

[1] Months of flowering [2] Distribution

DESCRIPTION	SCIENTIFIC NAME	COMMON NAME	M	D	HABIT
Heathland birchwood	*Anthoxanthum odoratum*	sweet vernal-grass	4-6	U	S
	Calluna vulgaris	ling, heather	7-9		W
	Deschampsia flexuosa	wavy hair-grass	6-7		S
	Galium saxatile	heath bedstraw	6-8		S
	Potentilla erecta	common tormentil	6-9	U	S
	Pteridium aquilinum	bracken	7-8	U	R
	Teucrium scorodonia	wood sage	7-9		R
	Trientalis europaea	chickweed winter-green	4-6	N	R
	Vaccinium myrtillis	bilberry, blaeberry, whortleberry, huckleberry	4-6		W

4. Ground layer

Mosses	*Hylocomium splendens*	glittering feather-moss			FM
	Plagiothecium undulatum	sharp fern-like feather-moss			FM
	Pleuozium schreberi	red-stemmed feather-moss		–	FM
	Polytrichum spp.	hair mosses			CM
	Rhytidiadelphus squarrosus	drooping-leaved feather-moss		U	FM
	Sphagnum spp.	bog mosses			CM

Liverworts, lichens, algae—uncommon

5. Fungi

	Amanita fulva	tawny grisette			CF
	Amanita muscaria	fly agaric			CF
	Fomes fomentarius				BF
	Lactarius torminosus	woolly milk-cap			CF
	Piptoporus betulinus				BF
	Russula nitida				CF
	Trametes spp.				BF

Pinewood

Pinewood has much in common with birchwood, for both are stages in colonization of recently treeless land. Pinewood is formed mainly on dry calcareous heathland. Pines are able to grow on such soils, partly because of the xeromorphic nature of their needle-like leaves. In Scotland, pinewoods are one of the natural or semi-natural types of vegetation, but in the south, where they occur on sandy soils, their establishment depends on the presence of nearby stands of planted pine trees to supply the wind-borne seeds in large quantities. Pinewoods usually have few other trees among them, though birch may be found in the gaps, and juniper in the more open parts of Scottish pinewoods. Larch may also be planted.

If the trees are close, the shading effect of pine is so great that few plants can survive beneath it. Pine is evergreen, so there is not even the early spring season of the deciduous woodland to allow growth of early-spring plants. Even the lower branches of pine cannot tolerate the shade, and these usually die off under close cover, leaving the pine trunks bare almost to the top. In the more open pinewoods, a number of low-growing shrubs may be found—probably remnants of the heathland in process of colonization. These have been classified in the list (p. 42) as 'shrub layer/upper-field layer', for, though woody and shrub-like, they merge with whatever field layer may be present. In fact, the field layer is usually very sparse, consisting of grasses and heath plants. Eventually, as the trees grow and light is cut off, the lower layers may die away altogether, leaving the soil covered with a deep layer of pine needles. These do not decay as readily as the leaves of deciduous trees, especially in the drier habitats of the sandy pinewoods in southern Britain. This thick dry layer of needles is an unsuitable medium for the germination of seeds, again contributing to the lack of a field layer.

Whereas with a birchwood the trees were eventually overgrown by beech or oak, the dense shade of pines precludes this, and pinewood is permanent.

The bark of pine flakes readily, so it makes a poor substrate for epiphytes. There are few lichens, mosses or algae on the trees, and these are found only in the better lit regions of the wood. In younger, more open woods, many mosses characteristic of heathland can be found below the field layer. As in beechwoods, fungi can grow in the dimly lit wood, obtaining nourishment from the leaf litter. Some typical species are included in the list (p. 43). *Boletus* is a cap-fungus, but instead of having its spores on gills, bears them in minute tubes, which open on to the underside of the cap. *Tricholomopsis rutilans* is found growing on old pine stumps, and *Fomes annosus* forms its brackets on the trunks of living pine trees. Dead branches and twigs are often attacked by *Leptosporus caesius.*

FLORA OF PINEWOOD

(for abbreviations, see endpapers)

DESCRIPTION	SCIENTIFIC NAME	COMMON NAME	M[1]	D[2]	HABIT
1. Tree layer					
Dominant	*Pinus sylvestris*	Scot's pine	5-6		W
Other trees	*Betula pubescens*	birch	4-5		W
	Juniperus communis	juniper	5-6		W
	Larix decidua	larch	5-6		W
	Populus tremula	aspen	2-3		W
	Sorbus aucuparia	mountain ash	5-6		W
2. Shrub layer/Upper field layer					
	Calluna vulgaris	ling, heather	7-9		W
	Erica cinerea	bell-heather	7-9		W
	Sarothamnus scoparius	broom	5-6		W
	Vaccinium myrtillis	bilberry, blaeberry, whortleberry, huckleberry	4-6		W
	Vaccinium vitis-idaea	cowberry, red whortleberry	6-8	N	W
3. (Lower) Field layer					
	Anthoxanthum odoratum	sweet vernal-grass	4-6		S
	Deschampsia flexuosa	wavy hair-grass	6-7		S
	Galium saxatile	heath bedstraw	6-8		S
	Holcus mollis	creeping soft-grass	6-7		S
	Potentilla erecta	common tormentil	6-9	U	S
	Veronica officinalis	common speedwell	5-8		S
4. Ground layer					
Liverworts	*Calypogeia trichomanis*				LLW
	Diplophyllum albicans				LLW
	Lepidozia reptans				LLW
	Lophocolea cuspidata				LLW

[1]Months of flowering [2]Distribution

DESCRIPTION	SCIENTIFIC NAME	COMMON NAME	M	D	HABIT
Mosses	*Campylopus flexuosus*				CM
	Dicranella heteromalla	silky fork-moss		U	CM
	Dicranum scoparium	lesser fork-moss			CM
	Hylocomium splendens	glittering feather-moss		+	FM
	Mnium hornum	swan's-neck thread-moss		U	CM
	Mnium punctatum	dotted thread-moss			CM
	Plagiothecium denticulatum	sharp fern-like feather-moss			FM
	Pleurozium schreberi	red-stemmed feather-moss		−	FM
	Pohlia nutans	silky pendulous thread-moss			CM
	Polytrichum commune	common hair-moss			CM
	Sphagnum spp.	bog mosses			CM
Lichens	few				
Algae	*Pleurococcus vulgaris*				
5. Fungi					
	Boletus spp.				CF
	Clitocybe spp.				CF
	Fomes annosus				BF
	Lactarius spp.	milk caps			CF
	Leptoporus caesius				BF
	Russula spp.				CF
	Tricholomopsis rutilans				

Alder carr

This is described on p. 60.

Mixed woods

Many woods do not conform strictly to the types described above. They may have had a varied history of felling and planting, and within the wooded area there may be variations of soil conditions that differentially favour certain species of tree. Their ecology is thus less simple than that of the woods dealt with so far, yet the same general principles can be applied. In many ways these are more satisfactory places for field study since different parts of the same wood can be studied separately and the results compared. By making contrasts between these areas, in which some conditions are identical while others differ, we can illustrate ecological ideas without having to travel farther than from one part of the wood to another.

For example, Stoke Woods, on a north-west facing slope of the River Exe, is a damp oakwood (*Quercus robur* dominant) with *Betula pendula*, *Acer campestre* and *Fagus sylvatica* as frequent components of the tree layer. Thus, this is not a typical damp oakwood, but could be called mixed deciduous woodland. In certain parts of the wood there were stands of *Pinus sylvestris*, giving a sharp contrast worthy of investigation. Similar stands of pine were scattered throughout the wood, so it seemed likely that they had been planted there, and had not become established because of local differences of soil composition. The variations in the vegetation beneath could probably be explained by the shading and other effects due to the two types of tree. Quantitative investigations were begun by a school party working in two areas of the wood—one in a mixed deciduous area, and one in a pine stand. Measurements showed that the pines were on average taller (30 m) than the deciduous trees (18 m), and were on average closer together (3 m apart compared with 3·5 m). In the shrub layer there were many seedlings of *Acer* and *Betula*. The number of seedlings of *Acer* decreased markedly between the time of a survey made in April and one made in July—had these been eaten, or were they less conspicuous among the taller, summer vegetation? Here was a chance to sort out a biotic factor, or look for a fault in technique. *Hedera helix* was a common member of the shrub layer in both areas, but *Rubus fruticosus*, though common in both, was dominant only beneath the pines. In the field layer, *Mercurialis perennis* was found only beneath deciduous trees, with *Asperula odorata* and *Geranium robertianum*, whereas *Pteridium aquilinum*, *Galeobdolon luteum* and *Betula pendula* (seedlings) were found only beneath pines. Taking the overall results of the survey it was noted that 28 different species were recorded in the field layer beneath deciduous trees, but only 11 different species beneath pine. Obviously, the latter conditions were more exacting. Perhaps the nature of the ground layer offered a clue, for beneath pine the soil was totally covered by leaf litter, whereas beneath deciduous trees leaf litter occupied only 80 per cent with 3 per cent covered by moss, and 18 per cent bare soil. The latter area would be more suitable for the growth of field-layer seedlings. Already many unanswered, and perhaps unanswerable questions had been raised; already there was material for many further investigations.

FRESHWATER HABITATS

Environmental factors

Woodland and a freshwater habitat are a good pair of habitats for early fieldwork, for they are both rich in living organisms yet differ from each other in so many ways. In the wood, the shading and protection provided by the trees produces an environment suited to the other woodland organisms, and shields them from the extremes of climate. In freshwater, the water itself provides this moderating influence, but to take advantage of its protection aquatic organisms need to be adapted to the special conditions the water imposes. The study of freshwater centres around modes of adaptation.

In a freshwater habitat the most important environmental factors are:

1. *Water.* This essential component of living organisms is usually plentiful in aquatic environments. Water has a buoying effect on the organisms that live immersed in it, and makes it less necessary for them to have rigid skeletal tissues. The immediate limpness of aquatic plants when removed from water demonstrates to what an extent they depend upon buoyancy for supporting their weight. A second consequence of the density of water compared with that of air is that organisms living in water find that it is more resistant when they move through it, or it moves past them. Water animals need to be streamlined if they are to be able to move rapidly. Water plants rely on their flexibility to avoid being broken and damaged by strong water currents. Submerged leaves are usually finely divided into filamentous leaflets, which offer little resistance to water flow. This has the added advantage of creating a relative increase in surface area through which gases may be exchanged and mineral ions absorbed. The continuous presence of water makes vegetative reproduction of plants less hazardous, for detached portions of plants can remain floating permanently in water and carry out their normal functions or, if they are to take root, do not suffer from water shortage during the critical early stages of rooting.

Among the simpler aquatic plants, especially the algae, we find many that reproduce vegetatively by simply *fragmenting*. The plant becomes divided into two or more pieces, each of which is able to lead a separate existence. Among the higher plants several reproduce by *turions*, which are specialized buds, formed at the onset of winter, and which become detached from the parent plant. They sink to the bottom of the pond or stream and overwinter there, developing to form a new plant next spring. Such plants are indicated in the list (pp. 51-53) by the abbreviation 'Tur'. The duckweeds (*Lemna* spp.) reproduce vegetatively by forming young thalli which bud from the periphery of the parent thallus and then break off and float away. In autumn, the whole plant sinks to the bottom of the pond, where it remains protected from winter freezing until spring.

2. *Light.* At the surface, and to a depth of a few metres, light intensity is usually adequate for plant growth. In open water, intensity can be high, though nearer the water's edge there may be shading from overhanging trees. Thus, light intensity is usually far higher than that met with in woodland. The chief limitation will be due to opaque matter such as silt suspended in the water. If this is present in quantity, light intensity will fall off steeply with increased depth, and water plants will be restricted to the shallower regions.

3. *Temperature.* Water has a higher specific heat than any other substance, and consequently it can absorb or emit large quantities of heat with the least possible alteration in its temperature. This buffering effect ensures that temperatures fluctuate little from night to day and only slightly from season to season. This is a prime advantage of water as a habitat.

4. *Oxygen and carbon dioxide.* Exchange of gases is one of the problems of aquatic life. Diffusion in solution is slower than diffusion in air. The quantity of oxygen available in a given volume of water is small. At 0 °C the volume of oxygen dissolved in 1000 cm^3 of water saturated with air is only 10 cm^3, and is even less at higher temperatures. An equal volume of dry air contains 209 cm^3 of oxygen. An animal may quickly deplete the water around it of dissolved oxygen. Similarly, plants may use up the available dissolved carbon dioxide, yet give off more oxygen than can diffuse away, causing the formation of bubbles of gas rich in oxygen. In moving water the circulation induced by the water currents continuously brings water from the depths to the surface, where exchange of gases with the atmosphere occurs. In shallow streams with rocks or waterfalls, the churning of the water promotes full aeration.

5. *Mineral ions.* These are dissolved in the water, or contained in the mud or rocks that make the bed. The amounts and kinds of mineral contained depends on where the water has come from, what types of rock it has passed over on its course, and what material has been added to it from other sources. If a stream or river has received drainage from agricultural land that has been well manured or has been heavily treated with fertilizer,

the mineral content of the water is increased. Sometimes the level of mineral ions may become *too* high, and the water organisms are adversely affected.

6. *The substrate, or bed.* In still or slowly flowing waters this usually consists of deposited silt or mud. In this, plants can root and animals can burrow. The nature of the silt depends on the rock from which it originated. The amount of silt depends on the rate of water flow both at the site of erosion and at the site of deposition, and also on whether the source rock is hard or is soft and easily eroded. Silt may have very different properties from the soils occurring on the stream banks. Variations in the thickness of silt deposits may occur at bends or bays in the river, and this affects the type of vegetation found. In rapidly flowing streams there is little deposition; nearly always the stream bed is being eroded; the bottom is rocky, and less suitable for plants to root in. Some algae encrust the rocks, and just a few species of moss and lichen are able to gain hold on the rocky substrate. Among the higher plants, *Isoetes*, *Littorella*, and *Lobelia* can root on the rocky bottoms of pools and lakes, but not if there is a strong water current. The commonest higher plants capable of inhabiting a rocky bottom where there is a strong current are *Ranunculus fluitans* and *Groenlandia densa* (p. 59).

7. *Water current.* The water velocity may vary from well over 1 m s^{-1} down to zero. A rapid rate of flow promotes aeration, leads to erosion of the bed and banks, and makes it difficult for plants and animals to gain a hold on the bed. At the other extreme, slow currents (less than 0·3 m s^{-1}) or still water restrict exchange of gases between the water and the atmosphere, promote deposition of silt, and cause no difficulty to plants rooted in the bed. Animals are able to float or swim in the water without losing their station or being deflected from their course.

Plant life in still or slowly flowing water

This section deals with life in ponds, slow flowing streams, slow flowing rivers, lakes, canals, and ditches. Much of what is said can also be applied to artificial habitats such as garden ponds, water tanks, aquaria, and partly blocked gutters.

In still water, even in ponds and lakes, there are local currents caused by wind, the movements of the larger animals, and by convection. These can stir up the silt and accumulating organic debris from the bottom, making the water turbid and reducing light intensity in the deeper parts. Temperature remains very steady always, but the buffering effect depends partly on the volume of the water, so that fluctuations in a small pond are greater than those in a lake. Supply of oxygen and carbon dioxide is poor and may often be a limiting factor. This is especially likely if a pond is overhung by trees. These shed their leaves into the water during autumn, and the decaying organic material that accumulates on the bottom promotes the activities of bacteria to such an extent that near the bottom the supply of dissolved oxygen is completely exhausted. Anaerobic bacteria can take over the decay under these conditions, and some blue-green algae are able to grow in the absence of oxygen, but other forms of life are inhibited. Such a pond is an unsuitable place in which to begin field-work.

Local variations of shade, light, temperature, composition of the bottom, and so on, have marked effects on the distribution of plants within a freshwater habitat, and these effects can be investigated by measuring the relevant factors and observing the distribution of the various species. In some instances it may be possible to devise experiments to test whatever hypotheses are put forward to explain the observed distributions.

The plants have been listed (pp. 51-53) under six headings, though some species (for example, *Polygonum amphibium*) appear under more than one heading since they are tolerant of a wider range of conditions, with a consequent variability of habit. The free-floating plants (List A) include many microscopic forms and simple forms, though the angiosperms are represented by a few species. The commonest is *Lemna minor*, which, because of its high rate of vegetative reproduction, can spread rapidly to cover the entire free surface of a pond or canal in a few weeks. Plants of this group are not affected by the depth of water below them, so can be found far from shore. Plants of List B are restricted to the regions of shore or shallows because below a certain water depth they would receive insufficient light. The limiting depth depends on the turbidity of the water. They are also affected by the presence of plants floating above them on the water surface, especially by a dense cover of *Lemna*. List B includes two species of *Chara*, which belongs to a small group of plants, the Charophyta, all of which live in fresh water. They reproduce sexually, the zygote being a resting stage which eventually germinates to give a new plant. Some species become encrusted with a chalky layer of calcium carbonate, and it is from this feature that the group gets its common name 'stoneworts'. All the plants in List B are filamentous, or have long and narrow or finely divided leaves. In silt-laden waters the presence of bottom-rooting plants consolidates the mud and reduces the velocity of water flow. This, in turn, causes the rate of deposition of silt to increase. As the silt accumulates, and as organic debris from the rooted plants accumulates with it, the level of the bottom rises, and it becomes possible for the bottom-rooting plants with floating leaves (List D) to succeed those of List B. This is the beginning of a succession which can eventually lead to marsh and fen, and this story is continued later (p. 60).

The stone-encrusting plants (List C) are all algae, able to gain hold on a smooth substrate where other plant types cannot. *Enteromorpha* is a multicellular alga in the form of flimsy tubes several millimetres in diameter, and up to 150 mm long. It is coloured green, and is attached to the rock by a small disc or holdfast at one end of the tube. It is common in canals and other slow-flowing waters. Light intensity at the bottom of deep waters is insufficient, so plants of List C are not found there.

Nearer to the bank than either of the previously listed plants we find the members of List D, those rooting in the mud, and having their leaves floating on the water surface. These are all flowering plants and include some common species. On the water surface their leaves receive full light intensity, and have stomata in their upper epidermis which allows exchange of gases with the atmosphere. The upper surface is waxy so that any rain falling on it or water splashed on to it will quickly run off. The underside of the leaves has a thin cuticle, making it possible for mineral ions to be absorbed into the leaf from the water below.

The leaves of plants growing in shallower water (List E) may vary in form according to whether they are submerged or aerial. In *Ranunculus aquatilis* (and in some other aquatic species of this genus) the contrast between the finely-dissected submerged leaves and the typical laminar aerial leaves is very marked. Not only are the submerged leaves dissected, but the segments of leaf radiate from the petiole in three dimensions. This makes a bushy structure with a large surface area that is in contact with a large *volume* of water.

Plants of List E are usually found growing among those of List D, for there is seldom a clear-cut zonation. As type follows type in a succession the earlier colonizers die out gradually and the more recent colonizers grow up between them. Similarly, on the landward borders of the region occupied by plants of List E, we see the gradual increase in numbers of plants of List F. Although their leaves are all aerial their lower stems and root systems are submerged, and the amount of submergence fluctuates from times of flood to times of drought. It is interesting to cut across the stems of these plants below water level. Many species have hollow stems, or a spongy cortex (aerenchyma), through which gases can diffuse between the aerial parts and the submerged parts of the plant.

Among the plants of Lists D, E, and F, which constitute the field layer of the swamp and marsh regions, we may find various mosses and liverworts. This ground layer is partly in shallow water and partly in marginal mud. Taking the flowering plants of the swamp, marsh, and waterside (waterside plants, pp. 53-57) and comparing them with the field-layer plants of woodland, one is struck by the absence of early-flowering species among the waterside plants. This feature, so much an advantage in woodland, has no special merit here. Another difference is that there are many more annual plants among the waterside species. The bare mud around a pond or lake—perhaps deposited only recently by flooding—makes an open area into which seeds of animals can be carried and where they can germinate without competition. This advantage is only short-lived for as tall vegetation becomes established—especially the reed-like plants of the marsh region—a situation akin to woodland is established, with the reeds equivalent to a shrub layer. Beneath these the annuals will gradually be replaced by perennials. It is only into the newly laid down regions, or perhaps the regions where tall vegetation has been trampled by cattle (or fishermen?), that the annuals can enter.

No fungi have been listed, for few of the more conspicuous ones are common in aquatic habitats. Most water fungi are the more primitive species, without large fruiting bodies. One of the commonest is *Saprolegnia*, which is found in greyish masses on decaying plant and animal remains. Some species of this fungus are parasitic and attack animals such as fish, especially where the skin has been damaged. It is sometimes found attacking fish in aquaria.

Waterside plants

This list (pp. 53-57) includes plants commonly found growing in damp situations close to water. It can be used in conjunction with either *plants of still or slowly flowing water* (pp. 51-53) or *plants of medium to fast flowing water* (p. 59). Some points about the waterside vegetation have been mentioned above, and there will be further discussion of this area under the section dealing with succession in slow-flowing water.

PLANTS OF STILL OR SLOWLY FLOWING WATER

(for abbreviations, see endpapers)

DESCRIPTION	SCIENTIFIC NAME	COMMON NAME	M[1]	D[2]	HABIT
A. Free-floating in or on the open water					
	Anabaena				BGA
	Azolla filiculoides	fern			S
	Bacillariophyceae	diatoms (various)			PP
	Chlamydomonas spp.				PP
	Desmidiaceae	desmids (various)			PP
	Elodea canadensis	Canadian pondweed	5-10		S/Tur
	Hottonia palustris	water violet			S
	Hydrocharis morsus-ranae	frog-bit	7-8		S
	Lemna minor	duckweed	6-7	S	see p. 48
	Microcystis spp.				BGA
	Nostoc spp.				BGA
	Riccia fluitans	floating crystal-wort			TLW
	Scenedesmus spp.				PP
	Spirogyra spp.				FA
	Ulothrix spp.				FA
	Utricularia vulgaris agg.	greater bladderwort	7-8		Tur
	Volvox spp.				PP
	Zygnaema spp.				FA
B. Rooted in mud, with leaves totally submerged					
	Callitriche stagnalis	starwort	5-9		a/S
	Chara aspera	stonewort			see p. 48
	Chara hispida	stonewort			see p. 48
	Cinclidotus fontinaloides				CM
	Drepanocladus fluitans				FM
	Groenlandia densa		5-9		R
	Hippuris vulgaris	mare's tail	6-7		R
	Isoetes lacustris	quillwort	5-7		S
	Littorella uniflora	shore-weed	6-8		S
	Lobelia dortmanna	water lobelia	7-8	N	S
	Myriophyllum spicatum	spiked water milfoil	6-7		R/Tur
	Nitella spp.	stonewort			see p. 48
	Sphagnum subsecundum (bog pools)				FM

[1] Months of flowering [2] Distribution

DESCRIPTION	SCIENTIFIC NAME	COMMON NAME	M	D	HABIT
C. Encrusting or attached to stones etc, totally submerged					
	Cladophora spp.				FA
	Enteromorpha spp.				see p. 48
	Nostoc spp.				BGA
	Oscillatoria spp.				BGA
D. Rooted in mud etc, leaves floating on water surface					
	Callitriche stagnalis	starwort	5-9		a/S
	Nuphar lutea	yellow water lily, brandy bottle	6-8	S	R
	Nymphaea alba	white water lily	7-8		R
	Polygonum amphibium	amphibious bistort	7-9		S
	Potamogeton crispus	pondweed	5-10		R/Tur
	Potamogeton natans	pondweed	5-9		R
E. Rooted in mud etc, some submerged leaves, some aerial leaves					
	Hippuris vulgaris	mare's tail	6-7		R
	Littorella uniflora	shore-weed	6-8		S
	Menyanthes trifoliata	buckbean, bogbean	5-7		R
	Myriophyllum spicatum	spiked water milfoil	6-7		R/Tur
	Polygonum amphibium	amphibious bistort	7-9		S
	Ranunculus aquatilis	common water crowfoot	5-8		S
F. Growing in mud in shallow water or at edge of water, all leaves aerial					
	Alisma plantago-aquatica	water plantain	6-8	S	S
	Apium nodiflorum	fool's water-cress	7-8	S	S
	Butomus umbellatus	flowering rush	7-9		R
	Glyceria fluitans	flote-grass	5-8	U	S
	Iris pseudacorus	yellow flag	5-7	U	R
	Juncus bufonius	toad rush	5-9	U	a
	Littorella uniflora	shore-weed	6-8		S
	Equisetum fluviatile	water horse-tail	6-7		R

DESCRIPTION	SCIENTIFIC NAME	COMMON NAME	M	D	HABIT
	Eupatorium cannabinum	hemp agrimony	7-9	S	S
	Mentha aquatica	water mint	7-10		S
	Phragmites communis	reed	8-9		R
	Polygonum amphibium	amphibious bistort	7-9		S
	Polygonum hydropiper	water-pepper	7-9		a
	Rumex hydrolapathum	great water dock	7-9		S
	Sagittaria sagittifolia	arrow-head	7-8	S	Tur
	Schoenoplectus lacustris	bullrush	6-7		R
	Sparganium erectum	bur-reed	6-8		R
	Typha latifolia	cat's-tail, great reedmace	6-7		S

WATERSIDE PLANTS

(for abbreviations, see endpapers)

DESCRIPTION	SCIENTIFIC NAME	COMMON NAME	M[1]	D[2]	HABIT
A. Algae					
	Pleurococcus vulgaris			U	
B. Mosses and liverworts					
	Amblystegium serpens				FM
	Blasia pusilla				TLW
	Brachythecium rivulare				FM
	Cinclidotus mucronatus				CM
	Conocephalum conicum	great scented liverwort			TLW
	Drepanocladus revolens				FM
	Eurhynchium riparoides	long-beaker water feather-moss			FM

[1] Months of flowering [2] Distribution

Plate 3. Village pond, Gloucestershire, May. Within this pond are many of the common pond species, including moorhens (A). Organic material for the support of the pond animals is provided by the photosynthetic activity of the pond plants, *Elodea* and *Cladophora*, and by the decomposition of fallen branches (B) and leaves from the overhanging sycamores (C). This pond is in a state of transition. Twenty years earlier the water surface extended almost to the road edge, and to the dry-stone walls. Then the pond was used for watering cattle from the neighbouring farms. The constant passage of cattle to the water's edge kept the margin of the pond clear of vegetation. Nothing could grow on the constantly trampled mud. With the installation of a piped water supply to the village, the pond was used less and less for watering cattle. A photograph taken four years earlier than this one showed that the muddy margin had become colonized by meadow plants, mainly grasses and buttercups. Four years ago the edge of the water was much nearer the road (broken line). It is obvious that it has receded several metres during the last four years. On the far side of the pond a stand of *Epilobium hirsutum* (D) has become established. This is commonly found growing in damp situations. There has been a gradual lowering of the average water level during the years, which might be due to the increasing amounts of water being transpired from the region of the pond margin by the encroaching vegetation. This has left a small 'beach' (E) which, being in heavy shade, has only sparse vegetation.

These aquatic and marshy habitats may be contrasted with some very dry microhabitats in the vicinity. The limestone walls support nothing but the lichen, *Xanthoria* (see Plate 7, facing p. 86). In the foreground, a pile of chippings (F) left over from road-mending provides another dry microhabitat.

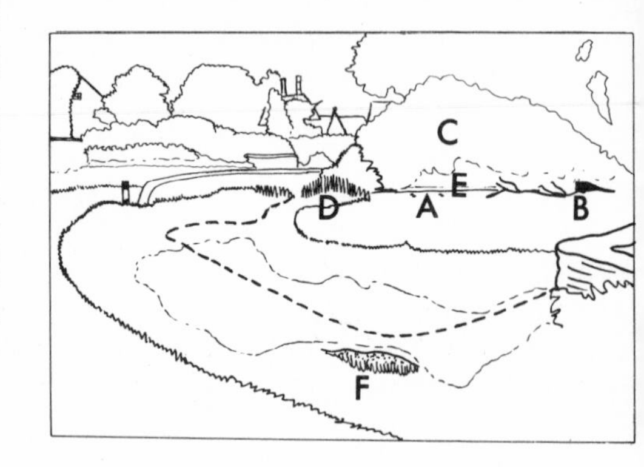

Plate 4. Pile of chippings beside the pond. Here rapid drainage produces conditions too dry for the wet-meadow plants, but *Potentilla anserina* (A, flowers black on diagram) easily becomes established, probably by seeds from plants on dry roadsides and farm tracks nearby. Whether this colonization will continue remains to be seen, and would make a subject for continued study. The grasses and creeping buttercup (*Ranunculus repens* — flowers white on diagram) might eventually spread on to the heap by vegetative means (rhizomes and and stolons) and overshadow the lower-growing *Potentilla*.

Plate 3. Village pond, Gloucestershire, May.

Plate 4. Pile of chippings beside the pond.

Plate 5. Hedgerow, Nottinghamshire, May.

Plate 6. Lower hedgebank and ditch of hedge shown above.

Plate 5. Hedgerow, Nottinghamshire, May. This shows the typical zones, from right to left:

- A Hedge—hawthorn, with ivy (G)
- B Hedge-bank—a variety of plants, including white dead nettle (*Lamium album*, H)
- C Water-filled ditch—containing fool's watercress (*Apium nodiflorum*, J)
- D Verge—with grasses, clovers, white dead nettle, and cow parsley (*Anthriscus sylvestris*, K)
- E Mown verge—on wide verges such as this only the part nearer the roadside will be mown. Here the mowings (L) have been left to rot. If this practice were to be continued over several years one would expect to find differences in vegetation and, possibly, in animal life in this region of the verge.
- F Verge edge—subject to hard wear from passing traffic (see p. 85)

For a discussion, see pages 82-86.

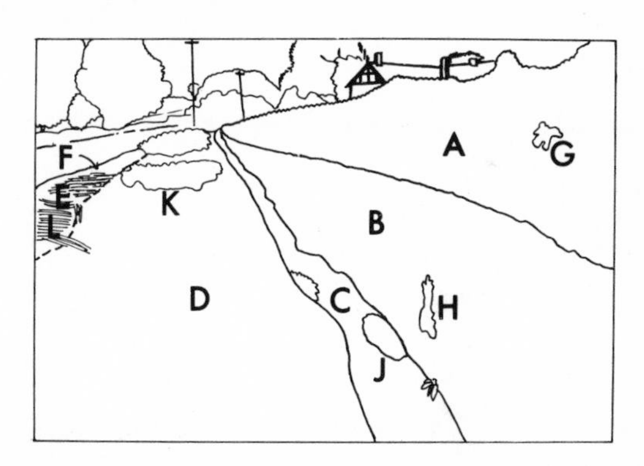

Plate 6. Lower hedgebank and ditch of hedge shown above. The hedgebank shows a variety of species including stinging nettle (A), goosegrass (B), bramble (C), Germander speedwell (D), sheep's sorrel (E) and white dead nettle (F). A few seedlings, mainly of speedwell, can be seen germinating at the lower margin of the hedgebank vegetation. There is a zone of bare soil below this—soil tests might help to explain why this region is not colonized. In the water, fewer plants are found. The soil surface below water level is covered with a thin film of algae (G), and the only flowering plant present is fool's watercress (*Apium nodiflorum*, H)

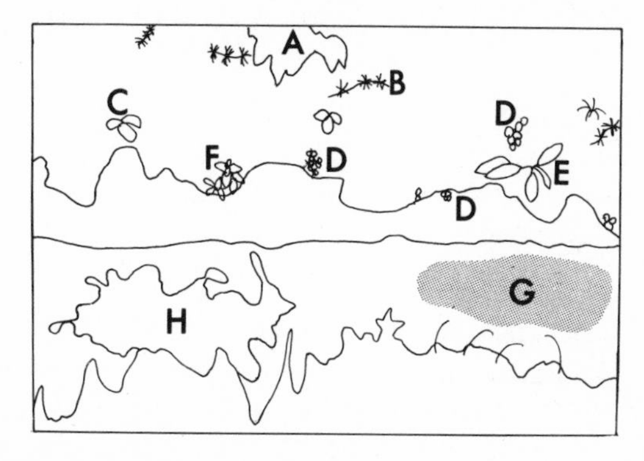

DESCRIPTION	SCIENTIFIC NAME	COMMON NAME	M	D	HABIT
	Leptodictyum riparium				FM
	Leskera polycarpa				FM
	Marchantia polymorpha	common liverwort			TLW
	Mnium longirostrum	long-beaked thread-moss			CM
	Pellia epiphylla	wide-nerved liverwort			TLW
	Preissia quadrata			N	TLW

C. Flowering plants (except trees)

DESCRIPTION	SCIENTIFIC NAME	COMMON NAME	M	D	HABIT
	Alopecurus geniculatus	marsh foxtail	6-7		S
	Armoracia rusticana	horse-radish	5-6	S	S
	Barbarea vulgaris	winter-cress, yellow rocket	5-8		b/S
	Caltha palustris	kingcup, marsh marigold	3-7	U	R
	Cardamine flexuosa	wood bitter-cress	4-9		S
	Cardamine pratensis	lady's-smock, cuckoo-flower	4-6	U	S
	Carex acutiformis	lesser pond-sedge	6-7		S
	Carex nigra	common sedge	5-7		S
	Chrysosplenium oppositifolium	opposite-leaved golden saxifrage	5-7		S
	Epilobium hirsutum	great hairy willow-herb, codlins-and-cream	7-8	S	S
	Epilobium parviflorum	small-flowered willow-herb	7-8		S
	Eupatorium canabinum	hemp agrimony	7-9	S	S
	Filipendula ulmaria	meadow-sweet	6-9	U	R
	Galium palustre	marsh bedstraw	6-7	U	S
	Geum rivale	water avens	5-9		R
	Hypericum tetrapterum	square-stemmed St John's wort	6-9		S
	Iris pseudacorus	yellow flag	5-7	U	R
	Juncus bufonius	toad rush	5-9	U	a
	Lycopus europaeus	gipsy-wort	6-9		R

DESCRIPTION	SCIENTIFIC NAME	COMMON NAME	M	D	HABIT
	Lysimachia vulgaris	yellow loosestrife	7-8		R
	Lythrum salicaria	purple loosestrife	6-8		S
	Mentha aquatica	water mint	7-10		S
	Montia fontana	blinks	5-10		a/S
	Myosotis scorpioides	water forget-me-not	5-9	U	S
	Phalaris arundinacea	reed-grass	6-7	U	R
	Polygonum amphibium	amphibious bistort	7-9		S
	Polygonum hydropiper	water-pepper	7-9		a
	Polygonum persicaria	persicaria	6-10	U	a
	Ranunculus ficaria	lesser celandine	3-5	U	T
	Ranunculus flammula	lesser spearwort	6-8	U	S
	Ranunculus sceleratus	celery-leaved crowfoot	5-9		a
	Sagina procumbens	procumbent pearl-wort	5-9	U	S
	Scrophularia aquatica	water betony	6-9	S	R
	Sonchus arvensis	field milk-thistle	7-10		S
	Sparganium erectum	bur-weed	6-8		R
	Stachys palustris	marsh woundwort	7-9		R
	Stellaria alsine	bog stitchwort	5-6		S
	Tussilago farfara	coltsfoot	3-4	U	S
	Veronica beccabunga	brooklime	5-9		S
D. Trees					
	Alnus glutinosa	alder	2-3	U	W
	Salix alba	white willow	4-5		W
	Salix cinerea	common sallow	3-4		W
	Salix fragilis	crack willow	4		W
	Salix viminalis	common osier	4		W
	Populus nigra	black poplar	4		W
	Ulmus glabra	wych elm	2-3		W

Plant life in rapidly moving water

Few plants can become established when the water velocity exceeds about 0·3 m s^{-1}. In the fastest streams, such as those in hilly areas, the velocity may be as great as 1·3 m s^{-1} and at this speed silt does not settle. The stream bottom is rocky, and no free-floating plants are found. The only vegetation comprises a few algae, mosses, and liverworts which manage to grow in crevices between the rocks, for in these crevices the water current is much reduced. Close to the rock surface, frictional effects reduce water velocity and, within about 4 mm of the rock surface, currents are usually so reduced that plants and animals are not washed away if they obtain a moderate grip on the substrate. A number of species can become established under these conditions and these are included in Lists B and C (p. 59). It is only further downstream or nearer the banks, where water velocity is less, that other larger plants can take root and remain undamaged by the water. These include the remaining species from List B; *Ranunculus fluitans* and *Groenlandia densa* both root between the rocks and have flexible shoots that trail downstream; *Enteromorpha intestinalis* has a similar habit.

At slightly lower water velocities the presence of bottom-rooted plants may retard the water flow further and lead to the deposition of limited amounts of silt. Here *Ranunculus aquatilis* can be found. Local configurations, such as bays, bends or hollows in the bed, may give rise to regions of low water velocity or even stagnation. In these regions a much wider variety of plants can grow, and many of these are the same as those found in slow-flowing water. List F is a guide to what is likely to be found.

If a stream or river is studied at several points along its course it is possible to chart the gradual change in vegetation along its length, from the very restricted number of species in its fast-flowing upper reaches, with the gradual addition of species as velocity diminishes and silt accumulates in its lower reaches. Eventually, velocity is diminished to that of slow-flowing rivers, when we find the vegetation characteristic of ponds and lakes (see pp. 47-53). The same gradation can sometimes be found across the width of a river. Velocity is greatest in the centre and decreases towards the banks, the vegetation correspondingly becoming more like that of still water.

PLANTS OF MEDIUM TO FAST-FLOWING WATER

(for abbreviations, see endpapers)

SCIENTIFIC NAME	COMMON NAME	M[1]	D[2]	HABIT
A. Free-floating in or on open water				
	None			
B. Rooted between rocks, with leaves totally submerged				
Callitriche stagnalis	starwort	5-9		a/S
Cinclidotus fontinaloides				CM
Fontinalis antipyretica	willow moss			FM
Groenlandia densa		5-9		R
Ranunculus fluitans	long-leaved water-crowfoot	6-8		R
Rorippa nasturtium-aquatica (= *Nasturtium officinale*)	water-cress	5-10		S
C. Encrusting or attached to rocks, totally submerged				
Cladophora spp.				FA
Enteromorpha intestinalis				see p. 48
D. Rooted between rocks, with some leaves floating on water surface				
Callitriche stagnalis	starwort	5-9		a/S
E. Rooted in mud etc, some submerged leaves, some aerial leaves				
Ranunculus aquatilis	common water crowfoot	5-8		S
F. Growing in mud in shallow water, or at edge of water, all leaves aerial				
	see pp. 52-53			

[1] Months of flowering [2] Distribution

Plant succession in slowly flowing water

This topic has already been touched on briefly (p. 48), and the story is resumed at the stage where the open-water plants (List A) and submerged plants (Lists B and C) are gradually giving way to rooted plants with floating leaves (List D). With stems and leaves occupying the water from bed to surface the velocity of water is even further reduced and the rate of silting even further increased. The dying remains of plants increase the amount of nutrient humus in the silty layer, making the growth of new plants even more abundant. Among these new plants are some from List E, able to grow in the shallower water that has resulted from silt accumulation. Since these can send some leaves above water level they shade the floating leaves of the earlier colonizers, which become reduced in numbers. Silt and humus accumulation proceeds rapidly in the water, which is now only a few centimetres deep. Soon the tall plants of List F appear, forming a reed swamp, dominated at first by *Typha latifolia* and later by *Phragmites communis.* In some localities the dominant reedswamp plant may be *Cladium mariscus.*

From the swamp stage the succession proceeds by the continued accumulation of silt and humus until the marsh stage is reached. Here the mud surface (we can now call it soil) is exposed, but the water level is only just below it, so the soil is waterlogged. It floods easily after heavy rains. From this stage the succession can follow one of several different sequences, depending on local conditions. The plants of this succession are listed on pp. 61-62. If water-movement is free, and if there is frequent flooding, the next plant to become established is usually *Carex paniculata.* This produces huge tussocks, among which *Carex acutiformis* and *Cladium mariscus* may grow. As the tussocks of C. *paniculata* increase in size, soil and debris accumulate upon them. Conditions on the tussocks are drier and less acid than in the soil below, and plants such as the marsh fern *Thelypteris palustris*, grow on the tussocks. The weight of the tussocks and the plants growing on them is too great for the loosely packed marshy silt to bear, and the tussocks repeatedly subside. The disturbance results in the formation of small pools of water on which open-water plants (List A, p. 51) appear. The submerged tussocks partly decay and give rise to large quantities of peat between the pools. Later, seedlings of *Salix* and *Alnus* become rooted in the tussocks, but with further subsidences the tussocks form a very unstable substrate and the young trees grow in a disorderly array with scarcely an upright trunk. In time, with the appearance of other plants and the gradual consolidation of peat and silt, conditions become more stable. The trees grow larger, and the beginning of the stage called swamp-carr is reached. Under the less marshy conditions many other plants can enter, including many waterside plants from list C (pp. 56-57). Tree seedlings become established, especially *Rhamnus catharticus* and *Frangula alnus*, which are soon followed by *Betula pubescens* and *Quercus robur* growing among the alders already present. The succession has now reached its final woodland stage, alder carr.

In regions in which water movement is less, usually farther from the river, and with less frequent flooding, the succession follows a slightly different route. Instead of *C. paniculata*, the dominant plant is *C. acutiformis*. This is a creeping sedge which does not form tussocks. It provides an even surface on which there are no conditions suitable for the germination of tree seeds. The sedge grows, accumulating more humus, to which is added more silt during the occasional floods, but this process is relatively slow, and gradually a firm and even layer of peat is built up. By the time trees are able to grow the substrate is able to support their weight. In this semi-swamp carr the tree trunks are more nearly vertical, and under the shade of the trees *C. acutiformis* is able to continue to grow along with the other carr species that gradually establish themselves there.

In some other regions of reed swamp, usually where water movement is least, the saw sedge, *Cladium mariscus*, becomes dominant. This sedge forms dense stands, crowding out all other species, and its remains are very resistant to decay, so helping to build up a firm substrate. This is eventually colonized by low trees and shrubs such as *Salix repens* and *Myrica gale*, which are followed by *Alnus*, *Betula*, *Quercus*, and the usual herbaceous species of alder carr.

Other modes of succession occur and these can often be worked out by local observations. The situations can be complicated at times, especially where drainage schemes have recently been altered and rivers have been dredged, so altering the rate of silt accumulation or the frequency of flooding. The examples given above demonstrate the principles of succession in water-side areas, and these are applicable to other combinations of conditions.

PLANTS OF THE SUCCESSION FROM REED SWAMP TO ALDER CARR

(for abbreviations, see endpapers)

DESCRIPTION	SCIENTIFIC NAME	COMMON NAME	M[1]	D[2]	HABIT
Trees	*Alnus glutinosa*	alder	2-3	U	W
	Betula pubescens	birch	4-5		W
	Frangula alnus	alder buckthorn	5-6	S	W
	Quercus robur	common oak, pedunculate oak	4-5	U	W
	Rhamnus catharticus	buckthorn	5-6	S	W
	Salix cinerea	common sallow	3-4		W

[1] Months of flowering [2] Distribution

DESCRIPTION	SCIENTIFIC NAME	COMMON NAME	M	D	HABIT
Shrubs	*Hedera helix*	ivy	9-11	U	W
	Myrica gale	bog myrtle, sweet gale	4-5	N	W
	Ribes nigrum	black currant	4-5		W
	Ribes sylvestre	red currant	4-5		W
	Ribes uva-crispa	gooseberry	3-5		W
	Rubus caesius	dewberry	6-9	S	W
	Rubus idaeus	raspberry	6-8	U	W
	Rubus fruticosus	blackberry, bramble	6-9	U	W
	Salix repens	creeping willow	4-5		WR
Other flowering plants	*Calystegia sepium*	bellbine, larger bindweed	7-9		R
	Carex acutiformis	lesser pond sedge	5-6	S	S
	Carex paniculata	panicled sedge	5-6		S
	Cladium mariscus	saw sedge	7-8		S
	Humulus lupulus	hop	7-8	S	S
	Phalaris arundinacea	reed-grass	6-7	U	R
	Phragmites communis	reed	8-9		R
	Solanum dulcamara	bittersweet, woody nightshade	6-9		W
	Symphytum officinale	comfrey	5-6		S
	Typha latifolia	cat's-tail, great reedmace	6-7	S	R
	Urtica dioica	stinging nettle	6-8	U	Ro
Ferns	*Athyrium filix-femina*	lady fern	7-8		R
	Thelypteris palustris	marsh fern	7-8	S	R
Mosses	*Acrocladium cuspidatum*	pointed bog feather-moss			FM
	Cratoneuron filicinum				FM
	Drepanocladus aduncus				FM
	Mnium affine				CM
	Sphagnum fimbriatum				CM
Liverworts	*Marchantia polymorpha*	common liverwort			TLW
	Trichocolea tomentella				LLW

Animal life in fresh water

An aquatic environment provides steady and favourable conditions for the growth of plants, and this, in turn, means that animal life is prolific and varied. The animals are usually the main attraction of freshwater studies. As in woodlands the variety of species is so great that it is impossible to give complete lists. Instead, the commonest species will be named in an analysis of the main microhabitats. In small volumes of fresh water such as ponds, conditions are less constant than in larger volumes. Occasional dry spells or periods of flooding may alter the water composition and favour the growth of different species. Thus, the animal population of a pond may change markedly, not only from one time of the year to another but also from year to year. An animal common one year may be absent in the following year, its place being taken by some other species. In most ponds the depth of water is seldom so great that any part of its bed is below the level at which plants can grow. Among these plants animals can live, and so in ponds we find a higher proportion of bottom-dwelling animals, and relatively fewer planktonic types. In lakes, the bottom dwellers are confined mainly to the shallower margins, and in the main body of water planktonic types are relatively more abundant. Rapidly moving water presents the same problems of anchorage for animals as it does for plants, for few animals are able to swim fast enough to maintain their station in a rapid water current. Small bottom-dwelling forms are at an advantage, because they can live between or beneath rocks, so escaping the main force of the current. Some, like the larvae of *Simulium*, the buffalo gnat, anchor themselves to rocks by many hooks on a basal pad, which they attach to a web spun on the rock surface. The caddis flies in rapidly flowing streams build cases of pebbles, which not only help to camouflage them against the stony bottom, and to a certain extent prevent them from being washed down stream, but also help to protect them from damage by loose stones that might dash against them in the turbulent water. Many animals that live on rocky bottoms of fast flowing streams are flat in shape, and this enables them to keep within the 4 mm layer of relatively slow moving water close to the rock surface. The larva of the mayfly, *Ecdyonurus venosus*, and the larva of the stonefly, *Dinocras cephalotes*, are both flattened, and so is the small fish, *Cottus gobio* (miller's thumb), which is thus able to live in the crevices between and under stones. Several other water animals have hooked claws that enable them to hang on to water plants.

In slow flowing or still water, conditions are favourable to a wider variety of animals, as suggested by the length of the list of common pond animals which follows. The list is arranged under various headings corresponding to separate or more-or-less separate regions of the pond. Note that, though most are found in ponds, some of the animals listed are not pond animals; this fact is indicated in the list where appropriate. The list for animals of rapidly-flowing streams (pp. 80-81), is relatively short. It will be seen that this is mainly due to there being so few microhabitats available in rapidly-flowing and turbulent water.

Studies in adaptation

The large numbers of animal species found in aquatic habitats make it possible to study adaptation with a wealth of examples. Details are given in most texts on freshwater biology, and the examples to be studied are best selected from those found at the site or sites under investigation. Adaptation studies can fall under the following headings.

1. *Methods of locomotion.* This topic can be restricted to locomotion in water, but it is interesting to make comparisons with methods of locomotion used on land and in the air. Two groups very suitable for study are the birds and the insects. In both these groups there are aquatic or partly aquatic terrestrial and aerial species, and their bodily adaptations to locomotion in these three environments show many points of contrast.

2. *Methods of obtaining oxygen.* If the study is restricted to aquatic animals there are still many different methods to compare; there are those animals that absorb oxygen from the atmosphere directly and those that take it from water. The structure of gills varies widely from group to group. A good introduction to this topic will be found in reference 40, p. 177.

3. *Methods of feeding.* Water animals provide many examples, and one could also include a study of the beaks and feet of water birds and waterside birds. The study of food and feeding methods can be linked with the study of food webs within an aquatic community, building up the web from organisms actually found at the site being studied. A web like that shown in Fig. 1 names groups rather than species, and makes a useful framework on which one can build up one's own web, using locally found species as examples. When worked out in detail, it is often found that the different species within a group have individual feeding preferences, and webs become very complex. With a study to this depth of detail usually only a part of the community can then be included (Fig. 2).

4. *Living in fast-flowing streams.* Some ideas can be got from p. 63.

Pollution

As human populations increase, and more waste materials need to be disposed of, pollution of our environment increases alarmingly. Pollution of fresh water is chiefly due to:

1. Drainage from heavily fertilized or heavily manured agricultural land. An excessive concentration of certain mineral ions may inhibit growth of plants.

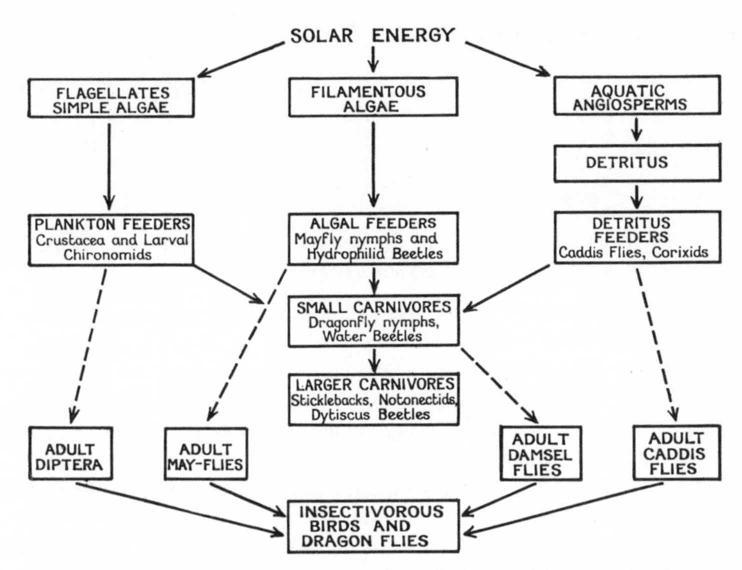

Fig. 1. Food-web of a freshwater habitat, Yellow Hill Delph, Blackburn, Lancashire (from POPHAM, see Book List p. 177, no. 40)

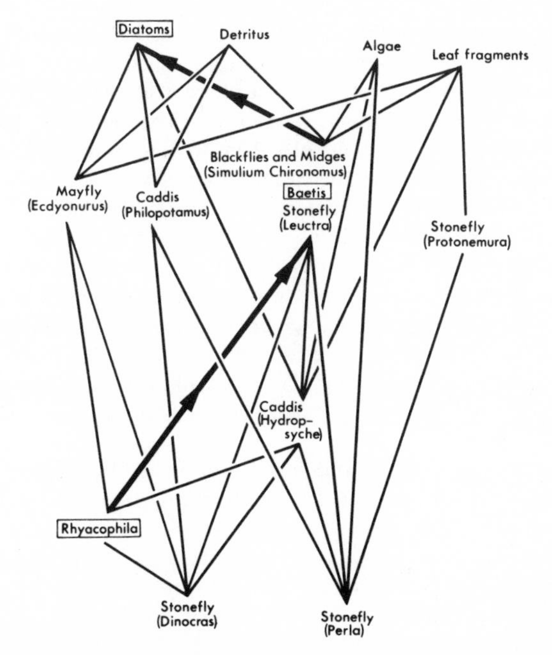

Fig. 2. Part of the food-web of a freshwater stream in Wales (redrawn from a diagram in PHILLIPSON, see Book List p. 177, no. 39, based on Jones, J.R.E. (1949): *J. Anim. Ecol.* 18, 142-159).

2. The spraying of ditches, ponds and occasionally larger areas of water, with pesticides, insecticides, or weed-killers.
3. Sewage.
4. Factory effluents.
5. Warm water used in cooling in various industrial processes, especially from the cooling towers of power stations.

Any of these can be investigated as a field project and the incidence of pollution correlated with changes in the plant and animal life present. Investigations are best made at several sites along a stream or river: above the site of pollution, immediately below it, and at various distances below.

Much pollution, including sewage, consists largely of organic materials. When these enter the water they promote the growth of bacteria to such an extent that the bacteria multiply rapidly, and soon the supply of dissolved oxygen is exhausted (compare with p. 46). In the anaerobic conditions created only anaerobic bacteria and a few water animals can survive. The surviving water animals are those that can breathe atmospheric oxygen, for example the rat-tailed maggot, *Eristalis*. When the breakdown of organic materials by the bacteria is well advanced, large quantities of mineral ions are released into the water and then the number of bacteria decreases. Further downstream, as oxygen concentration begins to rise again by absorption from the atmosphere, there is an increase in organisms (such as some protozoa) that feed on bacteria. A little further down there is an increase in the numbers of *Tubifex* and the larvae of *Chironomus*. The blood of these animals contains haemoglobin, which enables them to absorb adequate quantities of oxygen from the gradually increasing dissolved supply. The release of mineral ions into the water stimulates the growth of plants, especially algae, which by photosynthesis add more oxgyen to that already in solution. Eventually, at some distance from the site of pollution, and provided no further pollution has entered from a lower source, life returns more or less to normal, but the effects of pollution are drastic and some species may never reappear. This is understandable, for the freshwater food web is complex and any disturbance of it will have unpredictable and often permanent consequences. In such a finely balanced system any change is likely to be a change for the worse.

One type of organic material that is not destroyed by bacteria is that used as a detergent. Such materials are unaffected by sewage treatment and the action of the bacteria in the water. They persist, and even at a great distance from the site of pollution their presence in the water is made obvious by the foam that forms at waterfalls and weirs. Fortunately, this trouble is being avoided nowadays because detergents are being produced that can be decomposed by bacteria.

Thermal pollution causes depletion of the dissolved oxygen, for solubility of gases decreases with increased temperature. At the same time, the increase in temperature accelerates the metabolic rate of most water animals so that their oxygen requirements are greater. The higher temperature also reduces the affinity of haemoglobin for oxygen. The overall result is a severe oxygen shortage in the tissues of the animals. Other effects have been noted; the success of reproduction and the longevity of fish and other animals may be reduced. For example, *Daphnia* can live for 108 days if kept at 6 °C, but lives only for 29 days in water at 26 °C. Such changes in metabolic rates, reproductive rates, feeding rates, and oxygen requirements have diverse effects on the individuals and on the animal populations. Again, the delicate balance of the food web is threatened and the usual result is collapse of the system, leaving the habitat severely depleted of its normal stock of wild life.

FAUNA OF PONDS and of slowly flowing water

Abbreviations:

Stage of insects that show complete metamorphosis:

E = eggs
L = larvae
P = pupae
I = imagines (adults)

If all stages are spent in one location, no entry is made in this column.

Habitat

L = found in ponds and lakes
S = found in ponds and slowly flowing streams
R = found in ponds and slowly flowing rivers
C = found in ponds and canals
D = found in ponds and ditches
A = found in *or* on *or* over any still or slowly flowing waters
X = *not* found in ponds

(no entry signifies 'found only in ponds')

SCIENTIFIC NAME[1]	COMMON NAME	STAGE[2]	OTHER PLACES[3]	NOTES
A. Above the water				
Aedes spp.	mosquito	I		mating, laying eggs
Aeschna grandis	brown aeschna (dragonfly)	I I		
Anopheles spp.	mosquito	I	L	
Culex spp.	gnat	I		
Enallagma cyathigerum	common blue damselfly	I	S[4]	
Ephemera spp.	mayfly	I		
Ischnura elegans	common ischnura (dragonfly)	I	S	
Lestes sponsa	green lestes (dragonfly)	I	L	
Libellula quadrimaculata	four-spotted libellula (dragonfly)	I	L	

[1] Where no specific name given, several similar species may be found in a given location.
[2] Of animals that show metamorphosis. [3] If found elsewhere than in ponds.
[4] In this table the entry 'streams' refers only to slowly flowing streams; the same applies to rivers.

SCIENTIFIC NAME	COMMON NAME	STAGE	OTHER PLACES	NOTES
Myotis daubentoni	water bat, Daubenton's bat		A	feeding
Myotis nattereri	Natterer's bat		A	feeding
Pyrrhosoma nymphula	large red damselfly	I	S	mating, laying eggs

plus all the birds from list O, p. 79.

B. On the water surface—*above* the surface, supported by surface tension

Gerris lacustris	pond skater	I	SDL	
Gerris najas	pond skater	I	SDL	
Gyrinus natator	whirligig beetle	I	LS	
Hydrometra stagnorum	water measurer, water gnat	I	L	
Podura aquatica	springtail		L	
Velia caprai	water cricket		A	

C. On the water surface—*below* the surface film, supported by surface tension

Flatworms such as:

Planaria spp.	flatworm			

Insects such as:

Aedes spp.	mosquito	ELP		
Anopheles spp.	mosquito	ELP	L	
Culex spp.	gnat	ELP		

Molluscs such as:

Limnaea spp.	pond snail			
Planorbis spp.	ramshorn snail			

D. On the water surface—swimming or floating

Amphibia such as:

Bufo bufo	common toad		C	deep ponds
Bufo calamita	natterjack toad			shallow ponds
Rana temporaria	common frog		D	

SCIENTIFIC NAME	COMMON NAME	STAGE	OTHER PLACES[1]	NOTES
Birds such as:				
Anas crecca	teal		A	
Anas platyrhynchos	mallard			
Cygnus olor	mute swan		LSRX	
Cygnus cygnus	whooper swan		LSRX	
Fulica atra	coot		R	
Gallinula chloropus	moorhen		A	
Podiceps cristatus	great crested grebe		LX	
Podiceps ruficollis	dabchick, little grebe		RLX	
Mammals such as:				
Arvicola amphibius	water vole		RCX	
Lutra lutra	otter		RX	

E. Submerged—swimming or floating freely[1]

SCIENTIFIC NAME	COMMON NAME	STAGE	OTHER PLACES	NOTES
Protozoans such as:				
Actinophrys sol				
Chilomonas spp.				
Colpidium spp.				
Copromonas subtilis				
Euglena spp.				with chlorophyll
Lionotus spp.				
Paramecium spp.				
Stentor spp.				
Stylaria lacustris			LX	
Rotifers such as:				
Asplancha spp.			L	
Brachionus spp.			L	
Cathypna spp.				
Rattulus spp.			L	
Annelid worms such as:				
Herpobdella octoculata	leech		A	
Herpobdella testacea	leech		A	

1 Though some of the smaller species may spend the majority of their time attached to plants or rocks, they will be most often found swimming in a sample of water, having been dislodged by the process of taking the sample.

SCIENTIFIC NAME	COMMON NAME	STAGE	OTHER PLACES	NOTES
Roundworms (Nematode worms) such as:				
Gordius spp.	hairworm		L	
Crustaceans such as:				
Bosmina longirostris	cladoceran (water flea)		L	
Candona candida	ostracod		L	
Chirocephalus spp.	fairy shrimp			
Chydorus spp.	cladoceran (water flea)		L	
Cyclops spp.	copepod		A	
Cypris spp.	ostracod			
Daphnia spp.	cladoceran (water flea)		C	
Diaptomus spp.	copepod		L	surface waters
Eurycercus lamellatus	cladoceran (water flea)		CL	
Leptodora kindtii	cladoceran (water flea)		L	surface waters
Sida crystallina	cladoceran (water flea)			
Simocephalus vetulus	cladoceran (water flea)		LS	
Insects such as:				
Acilius sulcatus	water beetle			
Chaoborus plumicornis	phantom larva (of plumed gnat)	L	L	deep water
Chironomus spp.	bloodworm (larva of dance-gnat)	L		
Corixa spp.	lesser water boatman		L	
Dytiscus marginalis	great diving beetle		L	
Hydroporus ovatus	water beetle			
Hygrobia tarda	screech beetle			
Ilyocoris (= *Maucoris*) spp.	saucer bug		A	
Nepa cinerea	water scorpion		L	
Notonecta glauca	water boatman			edge

SCIENTIFIC NAME	COMMON NAME	STAGE	OTHER PLACES	NOTES
Philydris spp.	water beetle			
Ranatra linearis	water stick insect		L	
Arachnids such as:				
Limnochares aquatica	water mite			
Amphibia such as:				
Rana temporaria	common frog		D	
Triturus cristatus	great crested newt			
Triturus vulgaris	smooth newt			
Fish such as:				
Esox lucius	pike		LX	
Gasterosteus aculeatus	stickleback		LX	
Leuciscus rutilis	roach		LX	
Perca fluviatilis	perch		LX	
Phoxinus phoxinus	minnow		SX	
Salmo salar	salmon		RX	
Salmo trutta	trout		RX	
Mammals such as:				
Neomys fodiens	water shrew			
Algae such as:				
Chlamydomonas spp.				motile

F. On or among water plants, *above* water level

SCIENTIFIC NAME	COMMON NAME	STAGE	OTHER PLACES	NOTES
Insects such as:				
Sialis lutaria	alderfly	E	RLS	

and imagines of dragonflies, mayflies, damselflies and caddis flies (see lists A, L and M) *and* imagines of insects not primarily associated with an aquatic habitat.

SCIENTIFIC NAME	COMMON NAME	STAGE	OTHER PLACES	NOTES
Birds such as:				
Cygnus olor	mute swan			nesting on dense vegetation

SCIENTIFIC NAME	COMMON NAME	STAGE	OTHER PLACES	NOTES
Fulica atra	coot			nesting among reeds
Podiceps cristatus	great crested grebe			nesting among reeds

and animals not primarily associated with an aquatic habitat.

G. On or among water plants, on floating leaves

SCIENTIFIC NAME	COMMON NAME	STAGE	OTHER PLACES	NOTES
Coelenterates represented by:				
Hydra spp.			D	underside of leaves
Insects such as:				
Donacia spp.	water beetle	I	L	laying eggs on leaves
Nymphula nympheata	china mark moth			
Podura aquatica	springtail		L	
Molluscs such as:				
Planorbis spp.	ramshorn snail			underside of leaves

H. On or among water plants, below water level, on stems, leaves or roots; browsing, or attached to the plants

SCIENTIFIC NAME	COMMON NAME	STAGE	OTHER PLACES	NOTES
Protozoans such as:				
Arcella spp.				
Carchesium spp.				
Difflugia spp.				
Stentor spp.				
Vorticella spp.				
Sponges such as:				
Ephydatia fluviatilis	river sponge		L	encrusting plants
Euspongilla lacustris	pond sponge			encrusting plants

SCIENTIFIC NAME	COMMON NAME	STAGE	OTHER PLACES	NOTES
Coelenterates represented by:				
Hydra spp.			D	
Flatworms such as:				
Dalyellia spp.	rhabdocoel flatworm			
Dugesia lugubria			SL	
Rhynchomesostoma rostratum	rhabdocoel flatworm			
Annelid worms such as:				
Chaetogaster diaphanus	oligochaete		LX	
Glossiphonia complatata	leech		S	
Haemopsis sanguisuga	horse leech		S	
Herpobdella octoculata	leech		A	
Herpobdella testacea	leech		A	
Crustaceans such as:				
Sida crystallina	cladoceran (water flea)			
Simocephalus vetulus	cladoceran (water flea)		CL	
Insects such as:				
Corixa spp.	lesser water boatman		L	
Cyphon variabilis	beetle	L	L	floating vegetation
Haliplus spp.	water beetle		L	
Helochares lividus	water beetle			
Hydrobius fuscipes	water beetle		L	
Nepa cinerea	water scorpion		L	
Ranatra linearis	water stick insect		DL	
and many water beetles clinging temporarily to plants (see other lists).				
Arachnids such as:				
Argyroneta aquatica	water spider			
Hydrarachna spp.	water mite		SL	

SCIENTIFIC NAME	COMMON NAME	STAGE	OTHER PLACES	NOTES
Hygrobates spp.	water mite			
Neumania spp.	water mite			
Mediopsis spp.	water mite			
Molluscs such as:				
Aplector fontinalis	fountain bladder snail		SDX	
Bithynia tentaculata	common bithynia (gastropod)		CLX	
Hydrobia jenkinsi	Jenkin's spire shell		RC	
Limnaea palustris	marsh snail		A	
Limnaea stagnalis	common pond snail		A	
Planorbis albus	white ramshorn snail			
Planorbis complanatus	flat ramshorn snail		R	
Planorbis contortus	twisted ramshorn snail		S	
Planorbis corneus	great ramshorn snail		RL	
Planorbis planorbis	ramshorn snail		A	
Viviparus viviparus	freshwater winkle		RX	
Bryozoans (Moss animalcules) such as:				
Cristatella mucedo			L	
Lophopus crystalloides				
Plumatella repens			D	

I. Inside water plants

Eggs of *Notonecta*, and of some dragonflies; larvae of chironomid leaf miners.

J. On stones, on the bed of the pond, stream etc.

Annelid worms such as:				
Haemopsis sanguisuga	horse leech		S	
and other leeches.				

SCIENTIFIC NAME	COMMON NAME	STAGE	OTHER PLACES	NOTES
Molluscs such as:				
Limnaea stagnalis	common pond snail		A	
Bryozoans (moss animalcules) such as:				
Cristatella mucedo			L	

K. Beneath stones

SCIENTIFIC NAME	COMMON NAME	STAGE	OTHER PLACES	NOTES
Flatworms such as:				
Bdellocephala punctata			SL	
Dendrocoelum lacteum			L	
Polycelis nigra			L	
Insects such as:				
Baetis spp.	mayfly	L	RS	
Cloeon spp.	mayfly	L	DL	

L. On muddy bottom, or among bottom detritus

SCIENTIFIC NAME	COMMON NAME	STAGE	OTHER PLACES	NOTES
Protozoans such as:				
Arcella spp.				
Copromonas subtilis				
Difflugia spp.				
Stylonichia spp.				
Flatworms such as:				
Bdellocephala punctata				
Stenostomum spp.	rhabdocoel			
Roundworms (Nematode worms) such as:				
Dorylaimus stagnalis				
Rhabdolaimus aquaticus				
Annelid worms such as:				
Eiseniella tetrahedra	square-tailed worm		L	
Haemopsis sanguisuga	horse leech			
and other leeches.				

SCIENTIFIC NAME	COMMON NAME	STAGE	OTHER PLACES	NOTES
Crustaceans such as:				
Asellus spp.	water louse, pond slater		SL	
Astacus pallipes	river crayfish		RX	
Canthocamptus spp.	copepod		DL	
Gammarus pulex	freshwater shrimp		RSX	
Insects such as:				
Aeschna spp.	aeschna (dragonfly)	L	L	
Dicranota spp.	cranefly	L	S	
Enallagma cyathigerum	common blue damselfly	L	S	
Eristalis sp.	rat-tailed maggot (of a hoverfly)	L		stagnant polluted water
Hydroptila spp.	caddis fly	L	L	
Ischnura elegans	common ischnura (dragonfly)	L	CS	
Laccobius spp.	water beetle		L	
Leptocerus spp.	caddis fly	L	SLD	
Lestes sponsa	green lestes (dragonfly)	L	L	
Libellula quadrimaculata	four-spotted libellula (dragonfly)	L	L	
Limnophilus spp.	caddis fly	L	A	
Noterus clavicornus	caddis fly	L		
Oxyethira costalis	caddis fly	L	RSL	
Phryganea spp.	caddis fly	L	SL	
Pyrrhosoma nymphula	large red damselfly	L	S	
Sialis lutaria	alderfly	L	RSL	
Tabanus spp.	horse-fly	L	SL	
Tipula spp.	cranefly	L	S	
Trianodes spp.	caddis fly	L	L	
Arachnids such as:				
Limnochares aquaticus	water mite			
Macrobiotus macronyx	water bear			

SCIENTIFIC NAME	COMMON NAME	STAGE	OTHER PLACES	NOTES
Molluscs such as:				
Anodonta cygnea	swan mussel		LRX	
Limnaea stagnalis	common pond snail		A	
and other pond snails.				

M. In the mud, as burrowers or tube-dwellers

SCIENTIFIC NAME	COMMON NAME	STAGE	OTHER PLACES	NOTES
Annelid worms such as:				
Enchytraeus spp.	pot worm		A	
Lumbriculus variegatus	oligochaete		A	
Nais spp.	oligochaete			
Tubifex spp.	oligochaete			
Insects such as:				
Ephemera spp.	mayfly	L	LSX	
Sisyra spp.	spongilla fly	L		on sponges
Molluscs such as:				
Anodonta cygnea	swan mussel		RLX	
Dreissensia polymorpha	zebra mussel		RCX	
Pisidium amnicum	river pea-shell		RL	
Sphaerium corneum	horny orb-shell		SL	sand or gravel bed
Unio pictorum	Painter's mussel		RCLX	

N. Attached to water animals

SCIENTIFIC NAME	COMMON NAME	STAGE	OTHER PLACES	NOTES
Protozoa such as:				
Stentor spp.				on various animals
Vorticella spp.				on various animals
Annelid worms such as:				
Helobdella stagnalis	leech			on snails, etc.
Crustaceans such as:				
Argulus foliaceus	fish louse (copepod)			on fish and plankton
Arachnids such as:				
Hydrachna spp.	water mite			on water bugs

SCIENTIFIC NAME	COMMON NAME	OTHER PLACES	NOTES

O. Waterside animals

Many animals not primarily associated with water may be found above or beside freshwater habitats. A few of these are included in the list below and are marked *. They usually come to the water to feed on the insects which abound in the air above the water.

Amphibia such as:

SCIENTIFIC NAME	COMMON NAME	OTHER PLACES	NOTES
Bufo bufo	common toad	C	
Bufo calamita	natterjack toad		
Rana temporaria	common frog		
Triturus cristatus	great crested newt		
Triturus vulgaris	smooth newt		

Birds such as:

SCIENTIFIC NAME	COMMON NAME	OTHER PLACES	NOTES
Alcedo atthis	kingfisher	LR	
Anas crecca	teal	A	
Anas platyrhynchos	mallard	L	
Ardea cinerea	heron	LR	
Cygnus olor	mute swan	RLSX	
Cygnus cygnus	whooper swan	RLSX	
**Delichon urbica*	house martin	A	
Emberiza schoeniclus	reed bunting	A	
Fulica atra	coot	R	
Gallinula chloropus	moorhen	A	
**Hirundo rustica*	swallow	A	
Motacilla alba	white wagtail	S	
**Numenius arquata*	curlew	SL	
Riparia riparia	sand martin	RL	
Podiceps cristatus	great crested grebe	LX	
Podiceps ruficollis	dabchick, little grebe	RLX	

Mammals such as:

SCIENTIFIC NAME	COMMON NAME	OTHER PLACES	NOTES
Arvicola amphibius	water vole	A	
Neomys fodiens	water shrew		

Insects

Imagines of insects having aquatic larval stages, such as dragonflies, mayflies, damselflies, alderflies, midges, gnats, mosquitos, china mark moth. Water beetles in flight.

FAUNA OF RAPIDLY FLOWING WATER

(Abbreviation: L = larva)

A. Above the water

The list is similar to that for ponds, except that insect species listed there would be replaced by those which have larval stages adapted to life in rapidly flowing water (see lists J and K below).

B to I.

The disturbance of the surface, caused by rapid flow, makes it impossible for animals to remain on the surface by means of surface tension. The rapid streams are usually too fast for swimming animals, either on the surface or below it, either in open water or among water plants. The plants do not have floating leaves and there are very few plants to give protection to the animal population of fast-flowing waters.

J. On stones, on the bed of the stream

SCIENTIFIC NAME	COMMON NAME	STAGE
Flatworms such as:		
Polycelis felina		
Insects such as:		
Dinocras cephalotes	stonefly	L
Ecdyonurus spp.	mayfly	L
Simulium spp.	blackfly	L

K. Under stones

SCIENTIFIC NAME	COMMON NAME	STAGE
Flatworms such as:		
Polycelis felina		
Annelid worms such as:		
Eiseniella tetrahedra	square-tailed worm	
Insects such as:		
Agapetus spp.	caddis fly	L
Agriotypus spp.	ichneumon fly	L
Brachyptera risi	stonefly	L
Dinocras cephalotes	stonefly	L
Ecdyonurus spp.	mayfly	L
Helodes minuta	water beetle	L

SCIENTIFIC NAME	COMMON NAME	STAGE
Hydropsyche spp.	caddis fly	L
Isoperla spp.	stonefly	L
Leuctra spp.	stonefly	L
Nemoura spp.	stonefly	L
Philopotamus spp.	caddis fly	L
Plectrocnemia spp.	caddis fly	L
Rhyacophila spp.	caddis fly	L
Molluscs such as:		
Ancylastrum fluviatilis	river limpet	
Fish such as:		
Cottio gobio	miller's thumb	

L and M.

There is no muddy bottom in rapid streams, so no shelter for mud-dwelling animals. Occasionally one may find the bird *Cinclus cinclus* (dipper), walking and feeding on the bottom of fast streams.

O. Waterside animals

These may include several of the species listed on p. 79, plus animals characteristic of upland regions, such as *Motacilla cinerea* (grey wagtail).

HEDGEROWS

A glance at the flora lists (pp. 87-94) is enough to make it clear that hedgerows can harbour a great number of plant species. In consequence, the number of animal species is also large. Here, beneath the shade of the hedgerow bushes, grow many woodland plants, and in the damp ditch beside the hedge may be found waterside and aquatic plants. In a hedgerow, with its bushes, its hedge bank, its ditch, and a roadside verge, there are so many variations in environmental conditions that in small compass a hedgerow can provide a convenient and ideal habitat in which to investigate the effects of environment upon living organisms.

Hedgerow plants

The hedge bushes are usually planted closely, so that their branches firmly intertwine. A properly maintained hedge of hawthorn, the commonest farm hedge plant, has its longer, upward-growing shoots half severed at the base, and bent over sideways, to be woven among the branches in the body of the hedge. This gives even greater strength. Hedges should be trimmed regularly, for this promotes the development of the lateral buds and, hence, a compact and dense, bushy growth is produced. Thus, although the hedge bushes lack the stature of the woodland trees, the shade beneath them can be fairly heavy, especially to the north side, and this will encourage the growth of typical woodland species. This point will be considered again later.

The division of the list of woody plants into hedge bushes (List A) and occasional trees (List B) is to a certain extent arbitrary, since some of those in List A, if not trimmed during hedging, will grow to form trees, and some of those in List B may remain in bushy form in a trimmed hedge. In practice, if seedling trees establish themselves in a hedgerow they are cut out during hedging, for as they get bigger they will not form the close-knit system of branches that makes a hedge so effective a barrier. Occasional trees are planted or allowed to grow to full size from self-sown seedlings, and where these occur the local conditions approach even more closely those of woodland. It can be interesting to compare the vegetation of these shaded regions with that of open stretches of the hedgerow.

As in woodland, the bushes and trees provide a firm support for those plants with relatively weak but quick-growing stems—the climbers and scramblers. The distinction between climbers and scramblers is not always easy to make, but in general, a climber has twining stems, or tendrils, or adventitious roots, for anchoring itself permanently to its support. A scrambler is more haphazard, and relies only on thorns or hooked hairs to gain an often temporary grip upon its support. Examples of all these methods will be found in Lists C and D (pp. 87-88). Several are common woodland plants, and there are others not found in woodland, presumably because they require higher light intensity.

Hedges are usually planted on top of a raised hedge-bank, and on this bank special soil conditions may affect the growth of plants. The soil there may tend to be dry relative to soil nearby, mainly because of the ease with which soil water can drain from a raised bank. Banks that slope towards the south lose more water by evaporation, especially if they are exposed to direct sunlight. In dry conditions, plants of List E can become established. Many of these show special adaptation to dry conditions. Some have their leaves in a basal rosette. The rosette covers the ground closely, reducing further evaporation after the seedling has begun to grow and also making it difficult for other seedlings to grow in its vicinity. Very many of these plants are annuals. They germinate early in the year, before the warmer summer sunshine has had time to evaporate the soil water left from the snows and rains of winter. They grow, flower, and set seed in a short time, often in a few weeks, and their seeds remain in the soil ready to germinate early next year. Some of these plants (for example, *Galium aparine*) may also germinate in autumn, when soil moisture begins to increase again, but remain small, protected beneath the hedge and its leaf litter, until the following spring. Then they are already well established, and resume rapid growth. Not all plants in the list show clear-cut adaptations of form and life history. Working out the extent to which a given species is adapted to dry soil conditons can make a worth-while field project.

Local conditions may not be extreme, and may not be uniform, so it is possible to find species from Lists E and F growing side by side. Yet, those of List E are found more commonly on moister banks. The bank may be moist for one or more reasons:

1. It is shaded by the hedge, or by occasional trees.
2. It is near to a ditch containing water.
3. The soil level on the other side of the bank is more or less flush with the top of the bank, so that water draining from this soil runs over or through the bank.
4. The bank does not face towards the south.

In any of these situations, particularly the shaded, humid conditions brought into being by situation (1), the environment assumes many of the features of that of woodland. This helps to explain why so many woodland species occur in List F. There are many ferns, which need damp conditions for the growth of their prothalli and for sexual reproduction to occur. The plants of the shadier woodland areas (for example, *Mercurialis perennis*) do not appear in the list. Most of those listed are the plants characteristic of margins or open spaces in woods, and require a greater light intensity than the deeper shade plants.

Soil conditions may vary from top to bottom of the bank and, if the bank is high enough, distinct zones of vegetation can be observed. Lower down the bank, in the ditch itself, another set of conditions pertain. There is lower light intensity, greater shelter from wind, and higher relative humidity. Ditches differ in the amount of free water they contain, and this, too, can affect which plants grow there. Some ditches, because of alterations in the drainage system since they were dug, may seldom contain water. After exceptionally heavy rainfall they may fill for a day or two, but this water quickly drains away. Plants characteristic of these dry ditches are contained in List G. Some of these are small plants, found on most dry soils (*Potentilla anserina*, *Senecio vulgaris*, *Stellaria media*, and *Tussilago farfara*); the others on the list share the feature of being very robust plants, resistant to disturbance. This is a useful feature for a plant living in a ditch, for periodically the ditches are cleared by cutting down or digging out the plants. The only types able to re-establish themselves after such clearing will be annual plants, which can germinate afresh from seed, or tough perennials with deep stem stocks or root systems capable of vegetative reproduction. Ditch clearing and the weeding of garden beds have much the same impact on plants, and the plants found in dry ditches include many garden weeds as can be seen from List G and the list on pp. 99-102.

In damper ditches a wider range of plants can grow, again including many robust resistant forms. Here and on the verge (List J), which may be more or less continuous with the ditch bottom, continued cutting favours tough plants such as:

From list H	*From list J*
Arctium minus	*Agrimonia eupatoria*
Cirsium spp. (thistles)	*Anthriscus sylvestris*
Equisetum arvense	*Centaurea nigra*
Heracleum sphondylium	*Cirsium* spp. (thistles)
Stachys palustris	*Hieracium* spp. (hawkweeds)
Symphytum officinale	*Ononis repens*
Urtica dioica	*Pastinaca sativa*
	and the 5 grass species

The plants resist not only repeated cutting, but also trampling by cattle and horses, damage by cars parking on the verge, grazing by cattle herds, and other forms of rough treatment. They lack the protection provided by the proximity of the hedge and so must rely on stout stems and leaves, deep underground organs, thorns, or stinging hairs to allow them to survive.

Among this stout herbage a few quick-growing meadow plants are found, some of which are probably escapes from cultivation in farms, herb-gardens, or flower-gardens. These, like some of the herbaceous climbers from List D which are also found in damp ditches and on the verge, gain some support among the robust plants.

If the ditch constantly contains flowing fresh water, the growth of waterside and aquatic plants is encouraged (List I). The list includes only flowering plants, but several species of alga, moss or liverwort from the lists on pp. 51-57 could also be found in this situation. Usually the ditch will be part of a functioning drainage system and water flow will be slow, perhaps intermittent, but any silt that accumulates is removed when the ditch is cleaned, so there is no succession to swamp or marsh conditions and plants associated with this succession are rarely found in ditches. In ditches that are not properly maintained and allowed to fill in, there is some form of succession, depending largely on local circumstances.

The flora of the verge can vary considerably with the treatment the verge area receives. It may be mown regularly, it may be subject to the many hazards mentioned above, and in recent years there has been a disturbing tendency to add to this onslaught in the cause of efficiency by using herbicides to destroy the vegetation. Herbicides certainly seem to reduce the number of plant species to be found on a verge, for only the toughest species can survive. Against the short-term benefit of saving manpower must be balanced the long-term, probably irreversible effects of the destruction of one of our largest and most diverse reservoirs of wild plant and animal life. In recent years, too, we have begun to see some of the effects of the large-scale uprooting of hedges in order to make fields big enough for the economical use of combine-harvesters and other agricultural machinery. Already, deep drifts of soil, containing valuable seeds and fertilizers, have piled up on roads, having been blown from the fields when the protection against wind afforded by the hedges has been removed. Man has created deserts before, and hedge destruction is possibly one way of doing it. These are subjects that need more investigation, much of it within the scope of school biologists.

The impression may have been given that living conditions on the verge are mainly unfavourable, but there are some advantageous factors too. As distance from the hedge increases, light intensity increases, and these relatively open conditions favour many plant species. Further from the hedge there are fewer woodland and marginal woodland species. Another factor that favours the verge plants is that when ditches are cleaned the plant remains and silt are often dumped on the verge. In time, this material increases the mineral content of the soil and helps to produce lush verge vegetation.

Near the edge of the verge there is a sharp zone of transition between the moist, humus-rich soil and the arid, soil-less conditions of the tarmac or concrete road, with continual severe abrasion from passing traffic. At the edge of the verge a few plants are found able to survive occasional damage, gaining some benefit from the extra light available in this region (List K).

Two of those listed are grasses, a form of plant well able to withstand wear, and thus useful for lawns and playing fields.

On earth paths across the verge and on the sides of tracks to farm gates we find plants able to resist wear and to grow reasonably well in hard-compacted soils (List L). If, owing to the lie of the land, there are tracks and ruts that are frequently flooded and puddled, several species of *Juncus* are often found (List M).

The account shows what a wide range of environmental conditions exists around a hedgerow. The situation may be complex, and for this reason it is suggested that study of woodland and a freshwater habitat should if possible precede study of a hedgerow. After having seen woodland and freshwater factors in operation, the synthesis provided by hedgerow—and the additional microhabitats occurring there—may be more easily understood. One way of studying a hedgerow is to survey and contrast the plant and animal life on its two sides. Many differences will be found, and they may be ascribed to the effects of shading or variations in soil conditions. Differences of land use may also be apparent. On one side there is usually a verge, and on the other there may be a field of crops or grass. Grassland vegetation is described on pp. 128-136. Where a hedgerow abuts arable land there is usually at least a narrow gap between the hedge and the crop, and this may be explored. Here one may find the weeds of arable land (List N). The first two are extremely persistent weeds of cultivation; they have deep rhizomes that are almost impossible to remove, and that give rise to new plants by vegetative reproduction at a rapid rate. If attempts are made to dig up these plants, small portions of the rhizome easily break off and remain in the soil, to produce a new batch of plants soon after. These are two of the most troublesome garden weeds. The other plants on List N are all annual plants, well suited to the rapid colonization of the bare soil produced by ploughing, harrowing, and other forms of soil cultivation.

Plate 7. Old limestone wall, Gloucestershire, May. Two factors predominate in this microhabitat—the lack of water, and the calcium-rich nature of the stone. On more-or-less vertical surfaces only the lichen, *Xanthoria* (A), is found. The more horizontal stone surfaces have become covered with the silky wall feather moss, *Camptothecium sericeum* (B), another plant capable of resisting periods of extreme desiccation. From the larger crevices grow clumps of the succulent plant wall-pepper or biting stonecrop (*Sedum acre*, C). The moss and wall pepper are found only on top of the wall: this could be because of the greater rainfall or the greater illumination of this region. Perhaps both factors are important. Lower down the wall, though there are large accumulations of blown soil (shown black in the lower part of the diagram, within the dotted lines), few plants grow, even though seedlings (D) are capable of germinating there. Another plant found growing on this wall is herb Robert (*Geranium robertianum*), and some parts of the wall are completely covered with ivy. The commonest kinds of animal found on this wall are various species of spiders. The uneven surface of the wall provides prominences for the attachment of webs, which hang free across the crevices.

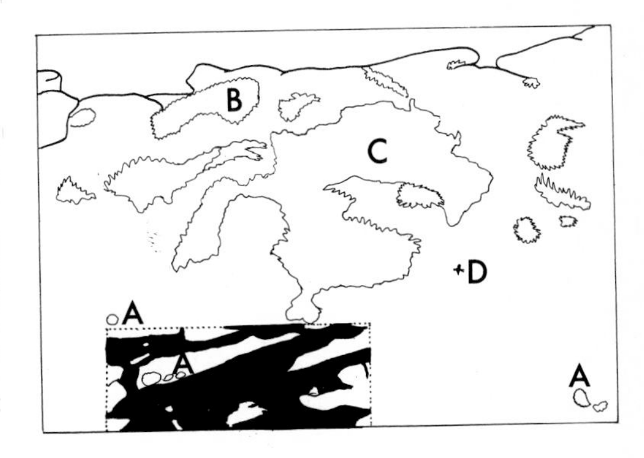

Plate 8. Limestone and cement parapet of bridge, Nottinghamshire, May. For a discussion see pages 105 to 106. In this photograph can be seen:

A (light stipple). *Pleurococcus*, on lower regions of painted rails (a splash zone?) and on horizontal limestone surfaces, especially around the bases of the rails, where water running down the rails would cause greater saturation of the stone. The white spots in this area are the remains of bird droppings.

B (medium stipple). Confluent colonies of *Lecidea*, occurring only on the stone, not on the cement.

C (dark stipple). White colonies of *Lecanora*, scattered along a zone that slopes down towards the right in the photograph. This zone seems to follow the cover provided by the vegetation growing in front of the wall. This lichen can be seen growing both on the limestone *and* on the cement-covered base (D). A few colonies of *Candellariella* occur just above this zone. Colonies of *Physcia* (black spots) found only on the sloping cement surface of the base.

E White dead nettle, rooted in crevices between the limestone blocks.

The vegetation in front of the wall, consists of typical verge plants: grasses, white dead nettle (F) and cow parsley (G). Hidden behind the vegetation in the position shown by the dotted outline (H) is a clump of the silky wall feather moss (*Camptothecium sericeum*), a moss commonly found on stone walls and apparently well adapted to such a habitat (see Plate 7). Here it is rooted in the crevice between two limestone blocks and in the cracks around the edges of the cement facing.

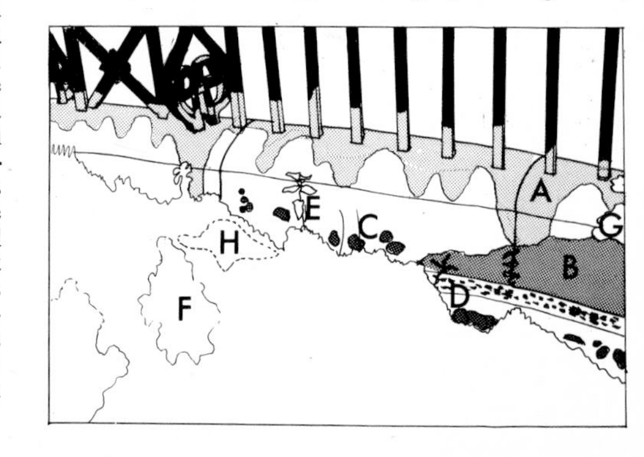

FLORA OF HEDGEROWS

(for abbreviations, see endpapers)

DESCRIPTION	SCIENTIFIC NAME	COMMON NAME	M[1]	D[2]	HABIT
A. Hedge bushes					
	Berberis vulgaris	barberry	5-6		W
	Carpinus betulus	hornbeam	4-5		W
	Corylus avellana	hazel	1-4	U	W
	Crataegus monogyna[3]	hawthorn	5-6	U	W
	Ilex aquifolium	holly	5-8		W
	Ligustrum vulgare	privet	6-7		W
	Prunus spinosa	blackthorn, sloe	3-5		W
	Ribes uva-crispa	gooseberry	3-5		W
	Sambucus nigra	elder	6-7	U	W
	Symphoricarpus rivularis	snowberry	6-9		W
	Thelycrania sanguinea	dogwood	6-7	S+	W
	Viburnum opulus	guelder rose	6-7		W
B. Occasional trees					
	Acer campestre	field maple	5-6	S	W
	Acer pseudoplatanus	sycamore	4-6	U	W
	Aesculus hippocastanum	horse chestnut	5-6	U	W
	Fraxinus excelsior	ash	4-5	U	W
	Malus sylvestris	crab-apple	5	S	W
	Prunus avium	gean, wild cherry	4-5		W
	Quercus robur	common oak, pedunculate oak	4-5	U	W
	Salix caprea	great sallow, goat willow	3-4		W
	Sorbus aucuparia	mountain ash	5-6		W
	Tilia × europaea	common lime	7		W
	Ulmus glabra	wych elm	2-3		W
	Ulmus procera	English elm	2-3		W
C. Woody climbers and scramblers					
	Clematis vitalba	traveller's joy, old man's beard	7-8	S+	W
	Hedera helix	ivy	9-11	U	W

[1] Months of flowering [2] Description [3] Commonest farm hedge bush

DESCRIPTION	SCIENTIFIC NAME	COMMON NAME	M	D	HABIT
	Lonicera periclymenum	honeysuckle	6-9	U	W
	Rosa arvensis	field rose	6-7	S	W
	Rosa canina	dog rose	6-7		W
	Rubus fruticosus	blackberry, bramble	6-9	U	W
	Solanum dulcamara	bittersweet, woody nightshade	6-9		W

D. Herbaceous climbers and scramblers

DESCRIPTION	SCIENTIFIC NAME	COMMON NAME	M	D	HABIT
	Bryonia dioica	white bryony	5-9	S	T
	Calystegia sepium	bellbine, larger bindweed	7-9		S
	Convolvulus arvensis	bindweed, cornbine	6-9	S	R
	Galium aparine	goosegrass, cleavers	6-8	U	a
	Galium mollugo	great hedge bedstraw	6-7		S
	Humulus lupulus	hop	7-8	S	S
	Tamus communis	black bryony	5-7		T
	Vicia cracca	tufted vetch	6-8	U	S
	Vicia hirsuta	hairy tare	5-8		a
	Vicia sepium	bush vetch	5-8	U	S

E. Plants of dry hedgebanks

DESCRIPTION	SCIENTIFIC NAME	COMMON NAME	M	D	HABIT
	Achillea millefolium	yarrow, milfoil	6-8	U	S
	Alliaria petiolata	hedge garlic, garlic mustard, Jack-by-the-hedge	4-6	U	b/R
	Arabidopsis thaliana	thale cress	4-5/9-10		a
	Bellis perennis	daisy	1-10	U	S
	Betonica officinalis	betony	6-9		R
	Briza media	quaking grass, doddering dillies	6-8		S
	Capsella bursa-pastoris	shepherd's purse	1-12	U	a
	Cerastium holosteoides (= *C. vulgatum*)	common mouse-ear chickweed	4-9	U	S
	Chelidonium majus	greater celandine	5-8	S	S
	Chrysanthemum parthenium	feverfew	7-8		S
	Clinopodium vulgare	wild basil	7-9	S+	R

DESCRIPTION	SCIENTIFIC NAME	COMMON NAME	M	D	HABIT
	Crepis capillaris	smooth hawk's-beard	6-9	U	a
	Festuca rubra	red fescue	5-7	U	S
	Galium aparine	goosegrass, cleavers	6-8	U	a
	Galium verum	lady's bedstraw	7-8	U	S
	Geranium molle	dove's-foot cranesbill	4-9		a
	Hieracium pilosella	mouse-ear hawkweed	8-10	U	R
	Hordeum murinum	wall barley	6-7	S	a
	Hypochaeris radicata	cat's-ear	6-9	U	S
	Parietaria diffusa	pellitory-of-the-wall	6-10	S	S
	Plantago lanceolata	ribwort	4-8	U	S
	Senecio jacobaea	ragwort	6-10	U	b/S
	Sisymbrium officinale	hedge mustard	6-7		a
	Smyrnium olusatrum	Alexanders	4-6		b
	Solidago virgaurea	golden rod	7-9		S
	Stellaria media	chickweed	1-12	U	a
	Taraxacum officinale	common dandelion	3-6	U	S
	Teucrium scorodonia	wood sage	7-9		R
	Verbascum thaspus	mullein, Aaron's rod	6-8		b
	Veronica chamaedrys	germander speedwell	3-7	U	S
	Veronica hederifolia	ivy-leaved speedwell	3-8	S	a
	Zerna ramosa	hairy brome	7-8		S

F. Plants of damp hedgebanks

DESCRIPTION	SCIENTIFIC NAME	COMMON NAME	M	D	HABIT
	Arum maculatum	lords-and-ladies, cuckoo-pint	4-5	S	T
	Asplenium adiantum-nigrum	black spleenwort	6-10		R
	Athyrium filix-femina	lady fern	7-8		R
	Brachypodium sylvaticum	slender false-brome	7		S
	Chaerophyllum temulentum	rough chervil	6-7		S
	Conium maculatum	hemlock	6-7		b
	Dipsacus fullonum	teasel	7-8	S	b
	Dryopteris dilatata	broad buckler-fern	7-9	U	R
	Dryopteris filix-mas agg.	male fern	7-8	U	R

DESCRIPTION	SCIENTIFIC NAME	COMMON NAME	M	D	HABIT
	Endymion non-scriptus	bluebell	4-6	U	B
	Festuca gigantea	tall brome	6-7		S
	Festuca pratensis	meadow fescue	6		S
	Galium palustre	marsh bedstraw	6-7	U	S
	Geranium dissectum	cut-leaved cranesbill	5-8		a
	Geranium robertianum	herb Robert	5-9	U	a
	Geum urbanum	wood avens	6-8	U	R
	Hypericum perforatum	common St John's wort	6-9		R
	Luzula pilosa	hairy woodrush	4-6		S
	Lysimachia nemorum	yellow pimpernel	5-9		S
	Oxalis acetosella	wood sorrel	4-5	U	R
	Phyllitis scolopendrium	hart's-tongue fern	7-8		R
	Polypodium vulgare	polypody	6-9		R
	Primula vulgaris	primrose	12-5		R
	Pteridium aquilinum	bracken	7-8	U	R
	Ranunculus ficaria	lesser celandine	3-5	U	T
	Ranunculus repens	creeping buttercup	5-8	U	S
	Saponaria officinalis	soapwort	7-9		R
	Silene alba	white campion	5-6		S
	Silene dioica	red campion	5-6	U	b/S
	Stachys sylvatica	hedge woundwort	7-8	U	R
	Stellaria holostea	greater stitchwort	4-6	U	S
	Viola odorata	sweet violet	2-4	S	R
	Viola riviniana	common violet	4-6	U	S
G. Plants in dry ditches					
	Ballota nigra	black horehound	6-10	S	R
	Lamium album	white deadnettle	5-12		R
	Lamium purpureum	red deadnettle	3-10	U	a
	Potentilla anserina	silverweed	6-8	U	S
	Senecio vulgaris	groundsel	1-12	U	a
	Stellaria media	chickweed	1-12	U	a
	Tussilago farfara	coltsfoot	3-4	U	S
H. Plants in damp ditches					
	Arctium minus	lesser burdock	7-9	U	b
	Armoracia rusticana	horse radish	5-6	S	S

DESCRIPTION	SCIENTIFIC NAME	COMMON MANE	M	D	HABIT
	Barbarea vulgaris	winter cress, yellow rocket	5-8		b/S
	Cirsium arvense	creeping thistle	7-9	U	Ro
	Cirsium palustre	marsh thistle	7-9	U	b
	Cirsium vulgare	spear thistle	7-10	U	b
	Epilobium montanum	broad-leaved willow-herb	6-8	U	S
	Equisetum arvense	common horsetail	4	U	R
	Galium palustre	marsh bedstraw	6-7	U	S
	Heracleum sphondylium	cow parsnip, hogweed, keck	7-8		S
	Polypodium vulgare agg.	polypody	6-9		R
	Stachys palustris	marsh woundwort	7-9		R
	Symphytum officinale	comfrey	5-6		S
	Urtica dioica	stinging nettle	6-8	U	Ro

I. Plants in ditches with constant fresh water

DESCRIPTION	SCIENTIFIC NAME	COMMON MANE	M	D	HABIT
	Apium nodiflorum	fool's watercress	7-8	S	S
	Callitriche stagnalis	starwort	5-9		a/S
	Caltha palustris	kingcup, marsh marigold	3-7	U	R
	Eleocharis palustris	common spike rush	5-7		S
	Filipendula ulmaria	meadow-sweet	6-9	U	R
	Myosotis scorpioides	water forget-me-not	5-9	U	S
	Myriophyllum spicatum	spiked water milfoil	6-7		R
	Oenanthe crocata	hemlock, water dropwort	6-7		T
	Potamogeton natans	pondweed	5-9		R
	Rorippa nasturtium-aquaticum	water-cress	5-10		S
	Scrophularia nodosa	figwort	6-9		R
	Veronica beccabunga	brooklime	5-9		S

J. Plants of the verge

DESCRIPTION	SCIENTIFIC NAME	COMMON MANE	M	D	HABIT
	Agrimonia eupatoria	common agrimony	6-8	S	S
	Anthriscus sylvestris	cow parsley, keck	4-6	U	S
	Artemisia vulgaris	mugwort	7-9		S

DESCRIPTION	SCIENTIFIC NAME	COMMON NAME	M	D	HABIT
	Centaurea nigra	lesser knapweed, hardheads	6-9	U	S
	Chrysanthemum (= *Tanacetum*) *vulgare*	tansy	7-9		S
	Cirsium arvense	creeping thistle	7-9	U	Ro
	Cirsium vulgare	spear thistle	7-10	U	b
	Cruciata chersonensis (= *Galium cruciata*)	crosswort, mugwort	5-6		S
	Cynosurus cristatus	crested dog's-tail	6-8	U	S
	Dactylis glomerata	cock's-foot	5-7	U	S
	Daucus carota	wild carrot	7-8		b
	Festuca ovina	sheep's fescue	5-8	U	S
	Hieracium spp.	hawkweeds	8-10	U	S
	Lapsana communis	nipplewort	7-9	U	a
	Lathyrus pratensis	meadow vetchling	5-8	U	
	Leontodon autumnalis	autumnal hawkbit	6-10	U	R
	Linaria vulgaris	yellow toadflax	7-10		R
	Lotus corniculatus	bird's-foot trefoil, eggs-and-bacon	6-9	U	S
	Lotus uliginosus	large bird's-foot trefoil	6-8		S
	Lysimachia nummularia	creeping Jenny	6-8	S	
	Matricaria matricarioides	pineapple-weed, rayless mayweed	6-7	U	a
	Medicago lupulina	black medick	4-8	U	a
	Myosotis arvensis	common forget-me-not	4-9		a
	Ononis repens	restharrow	6-9	+	R
	Pastinaca sativa	wild parsnip	7-8	S+	S
	Phleum pratense	Timothy	7	U	S
	Poa pratensis	smooth-stalked meadow-grass	5-7	U	R
	Potentilla anserina	silverweed	6-8	U	S
	Potentilla reptans	creeping cinquefoil	6-9	S	S
	Silene vulgaris	bladder campion	6-8		S
	Sonchus oleraceus	milk-thistle, sow-thistle	6-8	U	a
	Torilis japonica	upright hedge-parsley	7-8	U	a

DESCRIPTION	SCIENTIFIC NAME	COMMON NAME	M	D	HABIT
	Tragopogon pratensis	goat's-beard, Jack-go-to-bed-at-noon	6-7	S	a/S
	Trifolium arvense	hare's-foot	6-9		a
	Trifolium dubium	lesser yellow trefoil	5-10	U	a
	Trifolium hybridum	alsike clover	6-9		S
	Trifolium pratense	red clover	5-9	U	S
	Trifolium repens	white clover	6-9	U	S

K. Plants at the edge of the verge, by the road and paths

	Hordeum murinum	wall barley	6-7	S	a
	Lolium perenne	rye-grass	5-8		R
	Plantago lanceolata	ribwort	4-8	U	S

L. Plants on paths

	Bellis perennis	daisy	1-10	U	S
	Poa annua	annual poa	1-12	U	a/S
	Sagina apetala	pearlwort	5-8		a
	Sagina procumbens	procumbent pearl-wort	5-9	U	S

M. Plants of frequently flooded tracks, and cart-ruts

	Juncus articulatus	jointed rush	6-9	U	S
	Juncus bufonius	toad rush	5-9	U	a
	Juncus bulbosus	bulbous rush	6-9		S

N. Weeds of arable land

	Aegopodium podagraria	goutweed, bishop's weed, ground elder, herb gerard	5-7	U	R
	Agropyron repens	couch grass, twitch	6-9	U	R
	Capsella bursa-pastoris	shepherd's purse	1-12	U	a
	Papaver rhoeas	field poppy	6-8	S	a
	Cerastium glomeratum	sticky mouse-ear chickweed	4-9		a
	Galeopsis tetrahit	narrow-leaved hemp-nettle	7-9	U	a
	Raphanus raphanistrum	wild radish, white charlock	5-9		a

DESCRIPTION	SCIENTIFIC NAME	COMMON NAME	M	D	HABIT
	Senecio vulgaris	groundsel	1-12	U	a
	Silene vulgaris	bladder campion	6-8		S
	Sinapis alba	white mustard	6-8	+	a
	Sinapis arvensis	charlock, wild mustard	5-7	U	a
	Sisymbrium officinale	hedge mustard	6-7		a
	Spergula arvensis	spurrey	6-8	—	a
	Stellaria media	chickweed	1-12	U	a
	Thlaspi arvense	field penny-cress	5-7	S	a
	Veronica arvensis	wall speedwell	3-10	U	a
	Veronica hederifolia	ivy-leaved speedwell	3-8	S	a
	Veronica persica	large field speedwell	1-12		a

Hedgerow animals

When so many plant species are found growing in one place there is the opportunity for many animal species to take up residence. As in other habitats, there are hundreds of invertebrate species, mainly insects, associated with the plants. There are the leaf eaters, the leaf miners, the sap suckers, the pollen and nectar gatherers, the gall formers, and those that live in the leaf litter and surface layers of the soil. The plants of the verge and the hedge bushes provide ample support for the webs of spiders, and many small birds build their nests in the hedge bushes. Birds nesting in hedges include: wren, willow warbler, whitethroat, redstart, robin, sparrows, finches. A few larger birds (thrushes, blackbird) may also nest there, and occasional trees may provide nesting sites for other species.

If the ditch contains water we may find many of the usual freshwater animals. Among the large ones found in ditches are *Bufo bufo* (common toad) in deep ditches, and *Bufo calamita* (Natterjack toad) in shallower ditches. The common toad also spends time on the hedge-bank among the undergrowth and leaf litter. On many hedge banks the loose, crumbly nature of the soil, and the surface layer of leaf litter, covered by ivy and other plants, provides a safe refuge for many animals, and a breeding place for some. Among the smaller animals, wasps and solitary bees may build their combs in hollows in hedge-banks. Here one can find all the common British reptiles, *Anguis fragilis* (slow-worm), *Lacerta vivipera* (common lizard), and *Vipera berus* (viper, or adder).

Many small mammals live in or on hedge-banks, for not only is there protection, but the hedgerow provides a source of food in the form of berries, nuts, and insects. Also there may be crop plants in an adjoining field. *Erinaceus europaeus* (hedgehog) feeds on snails, slugs, insects and worms, all plentiful in a hedgerow, and conceals itself in the leaf litter of the hedge bank. Other hedgerow mammals include *Sorex araneus* (common shrew), *Sorex minutus* (pygmy shrew), *Mus musculus* (house mouse), and *Apodemus sylvatica* (long-tailed field mouse). The harvest mouse (*Micromys minutus*) builds its nest of a ball of grass supported between upright stems of grass and other plants. The doormouse (*Muscardinus avellanarus*) is commonly found in hedges, where it feeds on nuts, especially hazel, and builds its globular nest of twigs, moss and grass, supported in the hedge bushes. If the hedge-bank is dry and sunny, another inhabitant is the bank vole, *Clethrionomys glareolus*, which makes its nest of grass, moss, and feathers, supporting it in the hedge. Occasionally it inhabits a disused bird's nest.

Other organisms

So far there has been no mention of the simpler plants, and fungi, and little will be said about these. Mosses, liverworts, and lichens can be found in hedgerows, many species being the same as one would find in woodland, or in aquatic environments. Fungi are present in hedgerows, for example

Coprinus atramentarius (ink-cap) is commonly found on verges, but on the whole there are fewer conspicuous fungi here than in woodland. Those few that are found fit into the usual pattern of parasitism or saprophytism, but any serious study of the larger fungi is best done in woodland, where the variety is far greater.

PARKS AND GARDENS

Parks and gardens vary greatly in size, layout, and the extent to which they are cultivated. It is impossible to describe a typical garden or park for there are so many possibilities. Some may contain areas that are almost like woodland and may be treated as such for field studies (pp. 3-44). Many parks contain a pond or stream (pp. 45-81) and most parks and gardens have hedges (pp. 82-96) or walls (pp. 105-110). Thus, most of these other sections are also applicable to parks and gardens.

The more highly cultivated the garden, the less like one of the natural or semi-natural habitats it will be. The gardener tries, by cultivation, to favour certain selected 'garden' species (which he calls border plants, shrubbery plants or vegetables) at the expense of the wild species (which he calls weeds). In a highly cultivated garden it is interesting to study this contest between the gardener and his natural enemies, the weeds, the pests, and the diseases.

Plants of park and garden

The flora list (pp. 99-102) comprises mainly garden weeds, though a few other types are listed. It has not been found practicable to include all the hundreds of garden flowers and vegetables that are cultivated, but the list does include some of the more commonly planted trees. Even with trees the list would become excessively long if all the ornamental ones were included. Identification does present a problem here, but the better gardening handbooks are helpful, and a collection of illustrated flower-seed packets is easily made.

The London plane, *Platanus acerifolia*, is a tree very often planted in town and city parks, as well as in pavements. It seems able to survive the polluted urban atmosphere. This is probably due to its habit of flaking off small pieces of its bark frequently, so that deposits of soot and grime are soon got rid of. Its leaves have a smooth upper surface, so they are easily washed clean by rain, and it does not seem to suffer unduly by having its root system beneath paving or tarmac.

A garden bed is an area of soil cleared of plants on several occasions during the year, and usually planted with garden plants at a lower density than would be usual under natural conditions. Here is an ideal opportunity for weed plants to establish themselves. Not only is the soil relatively bare, but it has probably been brought to ideal condition and well supplied with just the right amount of water, humus, and mineral salts. It is an area ripe for colonization. The first weeds to colonize the bed have several features in common, which give them the ability to be the first to arrive and, having arrived, to stay. These features are indicated on the flora list, under 'habit'. A study of these features is a profitable exercise in a well-maintained (but not *too* well-maintained) garden. Usually, the first to become established are those that propagate rapidly by seed (habit G). There may be several reasons for rapid seed propagation:

1. rapid germination;
2. rapid growth and flowering;
3. large numbers of flowers on each plant;
4. large quantities of seed produced by each plant;
5. a long flowering period;
6. the seeds are ripe for germination as soon as they are released;
7. there is an ample source of seed because the plant is very common.

Some species show more than one of these features, and as a whole they are characteristic of annual plants. Features (1) and (2) are characteristic of the ephemeral annuals, which go through their entire life cycle in a few weeks, and then die leaving behind a plentiful supply of seeds. These can take advantage of very short periods favourable to growth; they are common in desert areas, for example.

For feature (7) the ample supply may arise because the plant is a widespread one, capable of living in many adjacent habitats. On the list many of the plants have universal distribution. Alternatively, the plant may be *locally* common and provide abundant seed because it has been introduced by the gardener himself. These are the self-sown garden plants (habit K), which once introduced into a garden become very difficult to eliminate. Another feature common to many annual weeds is the basal leaf rosette (habit J, described on p. 83).

If weeding of the bed is neglected, plants that reproduce very effectively by vegetative means (habit H) may arrive later. Once these weeds are established they crowd out the earlier annual weeds and are very difficult to eradicate. Many of them are deep-rooted (habit E) and cannot be removed by surface hoeing or by simply trying to pull them up. Such roots can be removed only by deep digging. Rhizomes of *Convolvulus arvensis* grow to a depth of 2 m, so considerable effort is needed to eliminate it. Even this may be unsuccessful, for many of these plants can reproduce themselves from pieces of stem or rhizome only 10 mm long (habit I), and it is usually impossible to extract every part of the plant from the soil. Such plants are among the most persistent weeds (category Z). Only a few plants have been marked I in the list, for information on this subject is scanty. Simple experiments can provide the information needed.

FLORA OF PARK AND GARDEN

(for abbreviations, see endpapers)

SCIENTIFIC NAME	COMMON NAME	M[1]	D[2]	HABIT	
Acer campestre	field maple	5-6	S	W	
Acer pseudoplatanus	sycamore	4-6	U	W	
Achillea millefolium	yarrow, milfoil	6-8	U	S	EHL
Aegopodium podagraria	goutweed, bishop's weed, ground elder, herb gerard	5-7	U	R	HZ
Aesculus hippocastanum	horse chestnut	5-6	U	W	
Aethusa cynapium	fool's parsley	7-8	S	a	
Agropyron repens	couch grass, twitch	6-9	U	R	EFHIZ
Agrostis canina	brown bent-grass, velvet bent-grass	6-7		S	L
Agrostis stolonifera	fiorin	7-8	U	S	LH
Agrostis tenuis	common bent-grass, browntop bent-grass	6-8	U	R	LH
Alliaria petiolata	hedge-garlic, mustard, Jack-by-the-hedge	4-6	U	b/S	E
Alyssum maritimum	sweet alyssum	6-9	G	a/S	GK
Anagallis arvensis	scarlet pimpernel	6-8	S	a/S	G
Antirrhinum majus	snapdragon	7-8	G	b/S	K
Atrichum undulatum	wavy-leaved thread-moss			CM	
Barbula spp.	beard mosses			CM	M
Bellis perennis	daisy	1-10	U	S	IJLMZ
Berberis vulgaris	barberry	5-6		W	
Betula pendula	silver birch	4-5		W	
Brachythecium rutabulum	rough-stalked feather-moss			FM	
Calendula officinalis	marigold	4-8	G	a	GK
Calystegia sepium	bell-bine, larger bindweed	7-9		S	EI
Capsella bursa-pastoris	shepherds' purse	1-12	U	a	GJZ
Cedrus libani	cedar of Lebanon	7	G	W	

[1] Months of flowering [2] Distribution

SCIENTIFIC NAME	COMMON NAME	M	D	HABIT	
Centaurea cyanus	cornflower	6-9	G	a	GK
Centranthus (= *Kentranthus*) spp.	red valerian	5-9	G	a/S	K
Cerastium holosteoides (= *C. vulgatum*)	common mouse-ear chickweed	4-9	U	S	FH
Ceratodon purpureus	purple-fruiting heath-moss			CM	
Chamaenerion angustifolium	rose-bay willow-herb	7-9	U	S	
Cheiranthus cheiri	wallflower	4-6	G	S	K
Chenopodium album	fat hen	7-10		a	G
Cirsium vulgare	spear thistle	7-10	U	b	E
Convolvulus arvensis	bindweed, cornbine	6-9	S	R	E
Cosmos bipinnatus	cosmos	7-9	G	a	K
Cynosurus cristatus	crested dog's tail	6-8	U	S	L
Dactylis glomerata	cocksfoot	5-7	U	S	
Dicranella heteromalla	silky fork-moss			CM	
Epilobium montanum	broad-leaved willow-herb	6-8	U	S	
Equisetum arvense	common horse-tail	4	U	R	EIZ
Eschscholtzia californica	Californian poppy	6-8	G	a	K
Euphorbia helioscopia	sun-spurge	5-10	S	a	G
Euphorbia peplus	petty-spurge	4-11	S	a	G
Eurhynchium praelongum	long trailing feather-moss			FM	
Festuca ovina	sheep's fescue	5-8	U	S	L
Festuca rubra	red fescue	5-7	U	S	L
Fraxinus excelsior	ash	4-5	U	W	
Fumaria officinalis	common fumitory	5-10		a	F
Funaria hygrometrica	common cord-moss				CM
Geranium dissectum	cut-leaved cranesbill	5-8		a	
Geranium molle	dove's-foot cranesbill	4-9		a	
Hedera helix	ivy	9-11	U	W	
Ilex aquifolium	holly	5-8		W	
Laburnum anagyroides	golden chain	5-6	G	W	
Lamium amplexicaule	henbit	4-8	+	a	FZ
Lamium purpureum	red dead nettle	3-10	U	a	
Larix decidua	larch	5-6		W	
Ligustrum vulgare	privet	6-7		W	
Lolium perenne	rye-grass	5-8		R	LM

SCIENTIFIC NAME	COMMON NAME	M	D	HABIT	
Lotus corniculatus	bird's-foot trefoil, eggs-and-bacon	6-9	U	S	
Lotus uliginosus	large bird's-foot trefoil	6-8		S	H
Lunularia cruciata	crescent-cup liverwort			TLW	
Matricaria matricarioides	pineapple-weed, rayless mayweed	6-7	U	a	GMZ
Mattiola spp.	stock	6-9	G	b	K
Myosotis arvensis	common forget-me-not	4-9		a	G
Myosotis, cultivated species	forget-me-not	4-6	G	S	K
Plantago lanceolata	ribwort	4-8	U	S	LMZ
Plantago major	great plantain	5-9	U	S	LMZ
Platanus acerifolia	London plane	5	G	W	
Poa annua	annual poa	1-12	U	a/S	GLM
Poa nemoralis	wood poa	6-8		S	L
Poa pratensis	smooth-stalked meadow-grass	5-7	U	S	L
Poa trivialis	rough-stalked meadow-grass	6	U	R	L
Polygonum aviculare agg.	knot-grass	7-10	U	a	M
Polygonum convolvulus	black bindweed	7-10		a	
Ranunculus ficaria	lesser celandine	3-5	U	T	H
Ranunculus repens	creeping buttercup	5-8	U	S	LHZ
Raphanus raphanistrum	wild radish	5-9		a	E
Sagina apetala	pearlwort	5-8		a	M
Sagina procumbens	procumbent pearlwort	5-9	U	S	M
Senecio vulgaris	groundsel	1-12	U	a	FGZ
Sinapis arvensis	charlock, wild mustard	5-7		a	
Solanum dulcamara	bittersweet, woody nightshade	6-9		W	
Sonchus asper	spiny sow-thistle	6-8	U	a	E
Sonchus oleraceus	milk-thistle, sow-thistle	6-8	U	a	E
Sorbus aucuparia	mountain ash	5-6		W	
Stellaria media	chickweed	1-12	U	a	FG
Tagetes erecta	African marigold	6-10	G	a	GK

SCIENTIFIC NAME	COMMON NAME	M	D	HABIT	
Taraxacum officinale	common dandelion	1-12	U	S	EIJLZ
Tilia × europaea	common lime	7		W	
Tortula muralis	wall screw moss			CM	
Trifolium dubium	lesser yellow trefoil	5-10	U	a	L
Trifolium repens	white clover	6-9	U	S	H
Tropaeolum majus	nasturtium	7-9	G	a	K
Tussilago farfara	coltsfoot	3-4	U	S	E
Urtica dioica	stinging nettle	6-8	U	Ro	EHZ
Veronica arvensis	wall speedwell	3-10	U	a	G
Veronica chamaedrys	germander speedwell	3-7	U	S	H
Veronica hederifolia	ivy-leaved speedwell	3-8	S	a	GZ
Veronica persica	large field speedwell, Buxbaum's speedwell	1-12		a	G
Veronica polita	grey speedwell	1-12		a	GZ
Viola arvensis	field pansy	4-10		a	

Plate 9. Waste land, Nottinghamshire, April. A very wide verge had been used as the site for dumping road metal during road repair work in the neighbourhood. This had eventually been cleared away, leaving a flat area, devoid of vegetation. Week-end motorists frequently parked here, so that the soil surface, covered with chippings, became compacted. Thus, a good parking site and picnic site became established in a part of the county offering relatively few other attractions to motorists. Continual wear from cars, the compact nature of the substrate and the complete absence of shade from trees or other tall vegetation have made this a difficult region for plants to colonize. Only a few species can stand up to such conditions, among them being greater plantain (*Plantago major*, A), with its rosette habit, rayless mayweed (*Matricaria matricarioides*, B) and lesser yellow trefoil (*Trifolium dubium*). What will be the future of this area? Straw (C) and other plant debris (light brown) is being blown from surrounding areas and is accumulating on the rough surface. In time this may decay and provide humus for better water retention and the nourishment of a more luxuriant cover of plants. This can happen only if people stop driving cars over the area and using it for picnics. On the other hand, as long as it is used by man, there is more chance of the accumulation of man's rubbish (D), much of it not subject to attack by bacteria of decay. A bleak future may be in store.

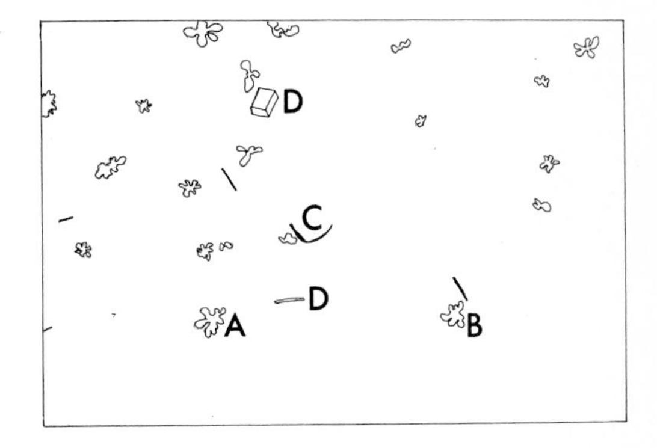

Plate 10. Pasture beside the River Idle, Nottinghamshire, April. This land is low-lying, with the water-table only just below the surface. After heavy rain it is flooded by overflow from the river. Under such waterlogged conditions certain grasses such as *Deschampsia caespitosa* are favoured and form dense tussocks among the pasture grasses as seen here. With intensive grazing by cattle or horses the *Deschampsia* is at a greater advantage, for though it is eaten by the animals it has tough leaves which are not so acceptable. The pasture grasses are closely grazed while the tussocks of *Deschampsia* are relatively unaffected. Such a pasture will gradually become less productive, but whether the land should be given improved drainage, ploughed and re-seeded, whether it should be turned over to some other purpose, or whether it should be left as waste land, depends more on economic factors, than on ecological ones. Ultimately, the consequences will be ecological.

Animals of park and garden

If conditions approach those of woodland, many woodland animals are present. Similarly, if there is a pond there are plenty of the smaller aquatic animals at least. The proximity of houses and other buildings increases the chances of finding animals associated with buildings, such as *Pipistrellus pipistrellus* (pipistrelle bat), *Myotis nattereri* (Natterer's bat), *Rattus rattus* (black rat), *Rattus norvegicus* (brown rat), and *Mus musculus* (house mouse). Several species of bird are common near habitation, including blue-tit, house sparrow, swallow, house martin, and robin. As well as these, almost any species can be found in a park or garden as an occasional visitor, especially if the area is secluded and if food is put out regularly.

The habits of the gardener, in growing many individuals of one species in a relatively confined area, results in an environment favourable for plant pests. It makes it easy for the pest to transfer from one individual to another of the same species. Pests attack the lush, cultivated plants from all possible quarters:

IN THE SOIL

eating roots, tubers, bulbs, rhizomes	insect larvae of	celery fly
		carrot fly
		crane fly (= leather-jackets)
		various moths (= cut-worms)
		click-beetle (=wire-worms)
		cockchafer
	millipedes	
	slugs	
	eelworms	
sucking sap from roots	lettuce aphid	

ON SOIL SURFACE

eating stems and leaves	rabbit
	birds
	tortoise

ON STEMS AND LEAVES

eating leaves	cabbage sawfly
	gooseberry sawfly
	cockchafer
	larvae of moths (winter moth on fruit bushes and rose trees; cut-worms leave soil at night and feed above ground)

	slugs (leave soil at night and feed above ground)
	snails
	larvae of cabbage white butterflies
	pea and bean weevils
	flea beetles (cabbage etc., wallflower)
sucking sap	aphids (broad bean, rose, apple, cabbage, etc.)
	woolly aphid (apple)
	capsid bugs (apple, fruit bushes, cherry, pear)
	frog-hopper
leaf miners	on holly, beet, celery, chrysanthemum

ON FLOWERS

aphids (roses)
bees
flies
butterflies } *collecting pollen or nectar* (NOT PESTS)*

ON FRUITS

apple sawfly
wasps (apples, pears, plums)

EATING SEEDS, *either on the plant, or in the soil where sown by the gardener.*

birds
mice
voles
larvae of pea moth

For further details of garden pests consult a gardening handbook. These animals, being dependent directly on the plants as a source of food, constitute part of the herbivore element of the garden food web. Their presence attracts to the garden a variety of carnivorous animals (centipedes, ladybird beetles, ants, spiders, insectivorous birds, birds of prey, and the domestic cat) which complete the food-web.

---

* This reminds us that many insects are not harmful, and the indiscriminate use of insecticides may do more harm than good. This is a gross interference with natural food-chains, for many birds depend on insects as a source of food, and so do some other animals.

WALLS

Walls may be studied as a microhabitat, or in conjunction with the study of a verge (pp. 85-86), a garden or park (pp. 97-104), an area of waste ground (pp. 111-115) or a meadow (pp. 129-132). Although they are essentially a very artificial habitat they have some features in common with the natural habitat of bare rocks or cliff faces so that there exist some species of plant able to colonize them.

The first colonizers are usually lichens and mosses, though the alga, *Pleurococcus*, may soon become established on a damp, shaded wall. A large number of these appear in the list (pp. 109-110), but not all are equally likely to be found. Some species, for example, grow well only on a medium with high calcium content. These correspond to those calcicolous plants that are chiefly found on chalky or limestone soils. The calcicolous wall plants (denoted +) are found almost exclusively on walls made of limestone or of cement. On brick walls they do not grow on the brick but on the mortar between the bricks. In very old walls, where material from the mortar has been leached into the surrounding brick, some of the species are found growing on the brick too.

Exposure to light, different conditions of humidity, and dampness of the wall material all affect the distribution of plants on the wall. Some are confined to wall top (denoted t), while others are found only near the base of the wall (*Madotheca platyphylla*, *Alliaria petiolata*). An interesting example of distribution of mosses and lichens was found on a wall in Nottinghamshire (see Plate 8). The wall ran north-to-south, and the east side of the wall was shaded heavily by trees. Nothing grew on that side. On its west side, open to full afternoon sunlight, growth was plentiful. The wall was low (about 0·5 m high) and topped by painted metal railings. The top surface of the wall, and the lower 5 to 8 cm of each railing were coated with *Pleurococcus*. Together, these formed the best-lit and dampest region of the wall, the lower ends of the railings presumably being wetted by splashing from the horizontal top surface. Water from the top surface drained down a vertical surface of limestone, of which the top of the wall was also composed. This surface had patches of *Pleurococcus*, especially just below the location of the thicker support railings where run-off of rain was greater in quantity. Elsewhere there were large white patches of the lichens *Lecanora calcarea* (+) and *Lecidea lucida*. There were also a few colonies of the lichen *Candel-*

lariella aurella (+) just above the zone of *Lecanora* colonies. The zonation of colonies tended to be nearer the ground at the south end of the wall, presumably because at this end the vegetation on the verge was shorter. The grass growing close to the wall provided shade and gave increased humidity lower down the wall. The top half of the wall was built of stone blocks, with crevices between them, in which the moss *Camptothecium sericeum* was growing. The bottom half of the wall was wider than the top half, and there was a shoulder sloping at an angle of about 45° on which grew a few colonies of *Physcia grisea* which was found nowhere else on the wall. On the vertical surface below this shoulder there were more colonies of *Lecanora* and *Candellariella*, but none of *Lecidea*. The absence of *Lecidea* was noted and, when an explanation was sought, it was soon found. On closer examination the lower half of the wall was discovered to be built of brick but faced with cement to make it resemble the stone of the upper half. Apparently *Lecidea*, which is not a calcicole, could survive on limestone but not on cement. Only two flowering plants grew on this wall. One was *Hedera helix*, invading from the adjacent verge, and the other was *Lamium album*, also present in abundance on the verge. A plant of this species had taken root in a large crevice where the cement had cracked away from the brickwork. *Lamium* is not mentioned in the list, for this illustrates the point that almost *any* species can live on a wall, especially if it is a locally common species and there is ample opportunity for seeds to become lodged in the wall. Provided the crevice is large enough and that enough wind-blown soil has accumulated, almost any seed will grow.

Though almost any species of herbaceous flowering plant can become established on a wall, at least temporarily, some species are better adapted to this habitat. They tend to be annual or biennial plants; presumably because in most walls there is little room for rhizomes, tubers, bulbs, and other bulky organs of perennation. Those that are perennials nearly always perennate by means of a long, narrow tap-root, which is ideally adapted to growing in wall crevices. In the list such plants are denoted by 'Tap'. The ivy-leaved toadflax (*Cymbalaria muralis*) shows an unusual adaptation to its habitat. Like many other wall species, its leaf-stalks and flower-stalks show a strongly positive phototropic response. This is to ensure that all leaves face towards the direction from which maximum light is received and that all flowers are easily visible to pollinating insects or accessible to the wind. In *Cymbalaria*, when a flower has been pollinated the response of its flower stalk is reversed and it becomes strongly negatively phototropic. The consequence of this is that the elongating flower stalk carries the ripening capsule towards the darkest regions, with an increasing chance of depositing the ripe capsule in a crevice where the seeds find ideal conditions for germinating.

Walls in different parts of the country may show differences in vegetation. There is a tendency for local stone to be used for wall building, so that limestone walls, with their calcicolous flora, tend to be found in limestone-yielding areas. Local climate can play a part, too; ferns are common on walls only in the dampest areas and in shady spots, and the list includes several flowering plants with restricted geographical distribution (N or S). *Umbilicus rupestris* has an extreme S distribution, being found only in south-west England and in Wales. *Aster tripolium* is even more restricted, being found in coastal areas and on sea-walls and being quite common in those places.

Wall studies do not usually lend themselves to work with animals. This is reasonable, for a wall is a bare habitat in the process of colonization. Until an adequate cover of plants has been established there is little to support an animal population. One can certainly find a few animals there, for there may be snails, slugs, and other small invertebrates browsing on the vegetation, and spiders may use the walls for supporting their webs. A number of dipterous insects (and others) also rest on walls during the day (for example, *Culex* species). When a wall becomes well overgrown with ivy or other climbing plants, or if a gardener has trained plants over the wall on wires or trellis, the possibilities for animal life increase. Small birds can nest or roost in the ivy, the abundant vegetation becomes the food source of the host of leaf-eating, leaf-mining and sap-sucking insects, and one may occasionally find a slow-worm or a common lizard in such situations. In spite of this occasional richness of fauna, a wall is not a community in the sense that a wood or a stream or pond can be considered to be a community. It is a microhabitat within a wood, a garden, or a derelict site, and within it there are usually several regions with different environmental conditions. It is in studying these factors and their interactions that most interest can be gained from working on a wall.

FLORA OF WALLS

(for abbreviations, see endpapers)

SCIENTIFIC NAME	COMMON NAME	M[1]	D[2]	HABIT
A. Flowering plants				
Alliaria petiolata	hedge garlic, garlic mustard, Jack-by-the-hedge	4-6	S	b/Tap
Antirrhinum majus	snapdragon	7-8	G	b/Tap
Arabidopsis thaliana	thale cress	4-5/9-10		a
Arenaria serpyllifolia	thyme-leaved sandwort	6-8		a/b
Aster tripolium	sea aster	7-10		b/S
Capsella bursa-pastoris	shepherd's purse	1-12	U	a
Cardamine hirsuta	hairy bitter-cress	4-8		a
Cerastium glomeratum	sticky mouse-ear chickweed	4-9		a
Cheiranthus cheiri	wallflower	4-6	G	Tap
Corydalis lutea	yellow corydalis	5-9		Tap
Crepis capillaris	smooth hawk's-beard	6-9	U	a
Cymbalaria muralis	ivy-leaved toadflax	5-9	S	S
Diplotaxis muralis	wall rocket, stinkweed	6-9	S+	a/Tap
Diplotaxis tenuifolia	perennial wall rocket	5-9	S	Tap
Draba muralis	wall whitlow-grass	4-5	+	a/b
Epilobium montanum	broad-leaved willow-herb	6-8	U	Tap
Erophila verna	spring whitlow-grass	3-6		a
Hedera helix	ivy	9-11	U	W
Hieracium pilosella	mouse-ear hawkweed	8-10	U	R
Lapsana communis	nipplewort	7-9	U	a
Mycelis (= *Lactuca*) *muralis*	wall lettuce	7-9	+	S
Parietaria diffusa	pellitory-of-the-wall	6-10	S	S
Parthenocirsus spp.	Virgina creeper	—	G	W
Sagina apetala	pearlwort	5-8		a
Sedum acre	wall-pepper, bitter stonecrop	6-7	+	S

[1] Months of flowering [2] Distribution

SCIENTIFIC NAME	COMMON NAME	M	D	HABIT
Senecio squalidus	Oxford ragwort	6-12	S	a
Reseda luteola	dyer's rocket, weld	6-8	S	b
Umbilicus rupestris	wall pennywort, navelwort	6-8	S	S
Veronica arvensis	wall speedwell	3-10	U	a

B. Ferns

SCIENTIFIC NAME	COMMON NAME	M	D	HABIT
Asplenium adiantum-nigrum	black spleenwort	6-10		R
Asplenium ruta-muraria	wall-rue	6-10		R
Asplenium trichomanes	maidenhair spleenwort	5-10		R
Ceterach officinarum	rusty-back fern	4-10	+	R
Cystopteris fragilis	brittle bladder-fern	7-8	N	R
Phyllitis scolopendrium	hart's-tongue fern	7-8		R

C. Mosses

SCIENTIFIC NAME	COMMON NAME	M	D	HABIT
Barbula convuluta	beard moss		Ttc	CM
Brachythecium rutabulum	rough-stalked feather-moss		U	FM
Bryum argenteum	silvery thread-moss		UTc	CM
Bryum caespiticium				CM
Bryum capillare	greater matted thread-moss		t	CM
Camptothecium sericeum	silky wall feather-moss		s	FM
Encalypta streptocarpa			+	CM
Encalypta vulgaris	common extinguisher moss		+	CM
Grimmia apocarpa			t	CM
Grimmia pulvinata	grey cushion-moss		t	CM
Hypnum cupressiforme	cypress-leaved feather-moss		U	FM
Orthotrichum duphinum			Ttc	CM
Rhacomitrum canescens			ts	CM
Tortula muralis	wall screw-moss		T	CM

SCIENTIFIC NAME	COMMON NAME	D	HABIT
D. Liverworts			
Lophocolea heterophylla		s	LLW
Madotheca platyphylla	flat-leaved scale-moss	+	LLW
E. Lichens			
Buellia canescens			LL
Caloplaca aurantica		+	CL
Caloplaca citrina		+	CL
Caloplaca heppiana		+	CL
Candellariella vitellina		+	CL
Cladonia fimbriata			LL
Cetraria glauca			LL
Hypogymnia physodes			LL
Lecanora calcarea		+	CL
Lecanora muralis		+	CL
Lecanora dispersa		UT+	CL
Lecidea coarctaca		Tt	CL
Lecidea lucida			CL
Parmelia saxatilis	crottle	U	LL
Physcia caesia			LL
Physcia grisea, etc			LL
Rinodina exigua			CL
Verrucaria nigrescens		+	CL
Xanthoria aureola			CL
Xanthoria parietina			CL
F. Alga			
Pleurococcus vulgaris		U	

WASTE LAND

For the purpose of this discussion, the term 'waste land' will be taken to mean land that has been disturbed once or repeatedly and then left relatively undisturbed, so that a new community is beginning to establish itself. The situation is one of colonization, such as has been described in connexion with parks and gardens (pp. 97-104) and walls (pp. 105-110). In those habitats colonization rarely proceeds far. All but the most dilatory gardeners will not allow too much re-establishment of natural vegetation, and the wall offers too restricted a habitat unless it crumbles and falls (when it ceases to be a wall). On waste land the early stages of colonization are followed by secondary stages in which the vegetation becomes dominated by the later arrivals, which are more vigorous herbaceous forms and which, in turn, give way partly to trees and shrubs. As succession proceeds, an increasing variety of animal life occupies the land.

Many species on the list (pp. 113-114) are the same as those listed as garden weeds and as wall plants, for in the early stages of colonization of waste land we find the same premium put upon rapid reproduction by seed. Those seeds that arrive first have the initial advantage. Wind dispersal is a very effective means for early entry, and plants that produce large amounts of wind-dispersed seeds or fruits (habit P) are frequent among the early colonizers (Q). Some of these, such as *Chamaenerion angustifolium*, follow their arrival by vigorous vegetative growth and perennation. This species is a frequent and permanent occupant of waste ground and was especially evident on the bombed sites in towns and cities after the Second World War. Other plants, such as *Senecio squalidus*, lose the advantage of their early arrival by being crowded out, for they are annuals and have no means of vegetative reproduction. Succession is a topic well worth studying in certain waste areas.

As well as the common weed species, a waste site may become colonized by escapes from cultivation. One of the commonest garden escapes to appear on waste ground is *Buddleia davidii*, a showy shrub with large lilac inflorescences. The name of this plant, butterfly bush, is well earned, for it is attractive to butterflies. In the summer, individuals of several species may be seen feeding at its flowers, even when few butterflies are seen on flowers around.

As time passes, a succession of plants appears on waste land and, if it is left undisturbed, seedlings of bushes and shrubs will begin to grow. This is less likely in areas that receive much trampling (where path-dwelling species (habit M) will remain dominant). If rabbits are abundant locally, the vegetation may become dominated by the rabbit-resistant plants, such as *Urtica dioica* and *Sambucus nigra*. The arrival of *Sambucus*, or in some localities *Ulex*, begins the succession to a scrubby vegetation.

FLORA OF WASTE LAND

(for abbreviations, see endpapers)

SCIENTIFIC NAME	COMMON NAME	M[1]	D[2]	HABIT	
Aegopodium podagrarius	goutweed, bishop's-weed, ground elder, herb gerard	5-7	U	R	H
Agropyron repens	couch-grass, twitch	6-9	U	R	H
Anthriscus sylvestris	cow-parsley, keck	4-6	U	S	O
Arctium minus	lesser burdock	7-9	U	b	O
Artemisia vulgaris	mugwort	7-9		S	
Atriplex patula	common orache	8-10		a	
Buddleia davidii	butterfly bush	6-7	T	W	
Capsella bursa-pastoris	shepherd's purse	1-12	U	a	
Cardamine hirsuta	hairy bitter-cress	4-8		a	NQ
Cerastium glomeratum	sticky mouse-ear chick-weed	4-9		a	Q
Chamaenerion angustifolium	rose-bay willow-herb	7-9	UT	S	HPQ
Coronopus squamatus	swine-cress, wart-cress	6-9	S	a/b	M
Crepis capillaris	smooth hawk's beard	6-9	U	a	PQ
Diplotaxis muralis	wall rocket, stinkweed	6-9	S+	a/Tap	Q
Geranium molle	dove's-foot cranesbill	4-9		a	N
Glechoma hederacea	ground ivy	3-5	U	S	H
Lactuca serriola	prickly lettuce	7-9	S	a/b	
Lapsana communis	nipplewort	7-9	U	a	O
Lamium album	white deadnettle	5-12		R	H
Lamium purpureum	red deadnettle	3-10	U	a	Q
Lotus corniculatus	bird's-foot trefoil, eggs-and-bacon	6-9	U	S	N
Malva sylvestris	common mallow	6-9	S	S	
Matricaria matricarioides	pineapple weed, rayless mayweed	6-7	U	a	M
Poa trivialis	rough-stalked meadow-grass	6	U	S	
Polygonum aviculare agg.	knot-grass	7-10	U	a	M
Potentilla anserina	silverweed	6-8	U	S	H
Potentilla reptans	creeping cinquefoil	6-9	S	S	H
Prunella vulgaris	self-heal	7-9		R	H
Ranunculus ficaria	lesser celandine	3-5	U	T	

[1] Months of flowering [2] Distribution

SCIENTIFIC NAME	COMMON NAME	M	D	HABIT	
Reseda luteola	dyer's rocket, weld	6-8	S	b	Q
Sagina procumbens	procumbent pearlwort	5-9	U	S	HM
Sambucus nigra	elder	6-7	U	W	O
Senecio jacobaea	ragwort	6-10	UT	b/S	PQ
Senecio squalidus	Oxford ragwort	6-12	ST	a	PQ
Senecio vulgaris	groundsel	1-12	UT	a	PQ
Silene vulgaris	bladder campion	6-8		S	
Stellaria media	chickweed	1-12	U	a	Q
Taraxacum officinale	common dandelion	1-12	U	S	PQ
Triplospermum maritimum (= *Matricaria maritima*)	scentless mayweed	7-9	U	a/S	
Tussilago farfara	coltsfoot	3-4	U	S	HP
Urtica dioica	stinging nettle	6-8	U	Ro	H
Veronica arvensis	wall speedwell	3-10	U	a	
Veronica chamaedrys	germander speedwell	3-7	U	S	
Veronica serpyllifolia	thyme-leaved speedwell	3-10	U	S	

It was said that succession rarely proceeds far in gardens, but occasionally a garden or lawn is left to run wild and can make an interesting study. It is not possible to give useful lists of plants that may be present, for so much depends on what was there to begin with. One would expect to find many of the weed plants from the garden list (pp. 99-102) and from the waste ground list (pp. 113-114). Later, as thicket or woodland conditions appear, woodland species can be expected. In spite of the gradual encroachment of the wild species, some cultivated species can hold their own and become truly naturalized. A derelict apple orchard in Huntingdonshire showed a characteristic open woodland vegetation, including *Urtica dioica*, *Stachys sylvestris*, *Heracleum sphondylium*, *Arctium minus*, *Rubus fruticosus* and many grasses, yet among these were well-established specimens that had either been planted there when the orchard was cultivated, or had escaped from the adjacent garden. These are listed below. In such areas it is as well to expect any type of plant.

INTRODUCED SPECIES FOUND NATURALIZED IN AN OVERGROWN ORCHARD

(for abbreviation, see endpapers)

SCIENTIFIC NAME	COMMON NAME	HABIT
Anemone ranunculoides	yellow wood anemone	R
Corydalis solida		a
Doronicum pardalianches	great leopard's-bane	T
Eranthis hyemalis	winter aconite	
Geranium endressii		R
Leucojum aestivum	summer snowflake, Loddon lily	B
Muscari atlanticum	grape hyacinth	B
Ornithogalum nutans	drooping star-of-Bethlehem	B
Prunus cerasifera	cherry-plum	W
Pulmonaria officinalis	lungwort	R
Tulipa sylvestris	wild tulip	B
Veronica filiformis		S

MOOR AND HEATH

It is not easy to give a perfectly satisfactory definition of either of these types of community, but in very general terms their features are:

Moor— on ancient rocks, with acid soils, in areas of heavy rainfall. Found chiefly in the north and west of Britain.

Heath— on light sandy soils, in areas of lower rainfall. Found chiefly in lowland England.

Moors and heaths exhibit extreme environmental conditions and consequently few plant species are well enough adapted to survive there. The vegetation tends to be dominated by one species over large areas, giving rise to a characteristic uniformity that makes these areas less interesting for prolonged study at school level. Although intermediates exist in profusion, the following main kinds of moor or heath can be distinguished.

Calluna moor

This is found mainly in areas of high rainfall and high altitude, such as the uplands of Scotland and parts of the Pennines. The soil is sandy and shallow, but does not drain easily because of the hard impervious layers beneath. These waterlogged conditions lead to oxygen deficiency in the soil atmosphere, with the result that the roots of plants are unable to function properly. The uptake of water is restricted, and is further reduced by the low soil temperatures that result from high altitude and the exposure of the damp soils to strong winds. Under these conditions of reduced water uptake, coupled with a tendency to excessive water loss from the leaves, only xeromorphic plants will be able to maintain themselves adequately. Of these, the dominant plant is *Calluna vulgaris* (heather). This shows many xeromorphic features such as the small, relatively thick leaves, with sunken stomata and thick cuticle, and is able to tolerate the acid soil conditions of the area. Under cold waterlogged conditions bacterial action is inhibited, so that decay of dead plant material is very slow and a thick layer of peat accumulates on and in the soil. In these sandy soils there is little calcium, and it is noticeable (see list on p. 118) that a high proportion of the common moorland plants are both xeromorphs (X) and calcifuges (—).

The slow rate of decay means that the normal cycles of use and re-use of mineral ions are not functioning fully. Most important is the deficiency of nitrogen in the soil. This puts at an advantage those plants that are able to acquire nitrogen from other sources. Leguminous plants such as *Ulex europaeus* have root nodules containing nitrogen-fixing bacteria. Others, such as *Calluna vulgaris*, *Vaccinium vitis-idaea*, *Erica* spp., and *Molinia caerulea*, rely on their association with species of fungi which grow in symbiosis with their roots, forming a mycorrhiza, and are able to fix atmospheric nitrogen. A few plants are able to obtain nitrogen from animal protein by trapping insects. Two examples included in the list are *Pinguicula vulgaris* and *Drosera rotundifolia*.

The vegetation varies from one part of the moor to another. The *Calluna*, which dominates most of the moor, is partly replaced in the drier areas by *Erica cinerea*, and in the wetter areas by *Erica tetralix*. In the wettest areas, where soil conditions are least favourable and where nitrogen deficiency is likely to be most severe, we find the insectivorous plants, and the mycorrhiza-bearing plant *Molinia caerulea*, as well as a few other species adapted to such extreme conditions. Man may also influence the distribution of vegetation deliberately or accidentally by means of fires. These burn the heath plants above ground, and heat the surface layers of soil so that shallow-rooted plants are completely destroyed, along with seeds buried in the surface layer. Only the deeper rooting plants can survive fires, and thus repeated firing favours *Calluna* and *Ulex*, which regenerate quickly from the living, buried stocks. They form a dense cover beneath which smaller species cannot become established.

FLORA OF CALLUNA MOOR

(for abbreviations, see endpapers)

DESCRIPTION	SCIENTIFIC NAME	COMMON NAME	M[1]	D[2]	HABIT
Dominant	*Calluna vulgaris*	heather, ling	7-9	–	XW
Dominant in drier areas	*Erica cinerea*	bell-heather	7-9	–	XW
Dominant in wetter areas	*Erica tetralix*	cross-leaved heath	5-6	–	XW
Other plants	*Campanula rotundifolia*	harebell	7-9		S
	Deschampsia flexuosa	wavy hair-grass	6-7	–	XS
	Empetrum nigrum	crowberry	5-6	N–	W
	Festuca ovina	sheep's fescue	5-8	U	S
	Galium saxatile	heath bedstraw	6-8		S
	Juncus squarrosus	heath rush	6-7	–	S
	Molinia caerulea	purple moor-grass	6-8	–	S
	Myrica gale	bog myrtle, sweet gale	4-5		W
	Nardus stricta	mat-grass	6-8		XR
	Potentilla erecta	common tormentil	6-9	U–	S
	Pteridium aquilinum	bracken	7-8	U–	XR
	Rubus fruticosus	blackberry, bramble	6-9	U	W
	Ulex europaeus	furze, gorse, whin	3-6	U	XW
	Vaccinium vitis-idaea	cowberry, red whortleberry	6-8	N	W
In dampest areas	*Carex* spp.	sedges		U	S
	Drosera rotundifolia	sundew	6-8	–	VS
	Eriophorum angustifolium	common cotton-grass	5-6	–	R
	Pinguicula vulgaris	common butterwort	5-7	–	V Rootless bud
	Sphagnum spp.	bog-moss		–	CM

[1] Months of flowering [2] Distribution

Vaccinium moor

This is formed in the same area as calluna moor, but mainly on the better drained hill-tops. The vegetation is very similar to that of calluna moor (pp. 116-118), except that the dominant plant here is *Vaccinium myrtillus* (bilberry, blaeberry, whortleberry, huckleberry, 4-6, —, XW).

Molinia-grass moor

In contrast to the calluna and vaccinium moors, which are on sandy soils, molinia-grass moor occurs on clay soils. Waterlogged clay soils bearing molinia moor are found in the Pennines, Wales, and Scotland. The dominant plant, *Molinia caerulea* (purple moor-grass), grows in dense tussocks, and is recognizable by its long, narrow, purple inflorescences in June, July, and August. This species is tolerant of the wet, acid soils that occur on the gentler slopes of moorland areas. On the upper slopes, which are usually steeper and therefore better drained, the *Molinia* is partly replaced by *Nardus stricta* (mat-grass). This can also be found near streams and flushes where conditions are less acid, and where the water movement helps to increase the quantity of oxygen available. At the other extreme, the level areas in the valleys may contain stagnant water, with very low oxygen concentrations. Under these conditions the dominant plant is usually *Eriophorum angustifolium* (cotton-grass, hare's tail). Thus, a molinia moor may contain areas of *Nardus* and of *Eriophorum*, depending on local variations in soil conditions. In some parts of the country larger tracts may be dominated by *Eriophorum* or by *Nardus*, forming cotton-grass moor and nardus-grass moor respectively (see pp. 121-123).

As with the calluna moor, the number of species that can tolerate the exposure to weather and the damp, acid soil conditions is limited, and the flora list is correspondingly short. Many of these are calcifuge plants, and a few exhibit xeromorphic features.

I

FLORA OF MOLINIA-GRASS MOOR

(for abbreviations, see endpapers)

DESCRIPTION	SCIENTIFIC NAME	COMMON NAME	M[1]	D[2]	HABIT
Dominant	*Molinia caerulea*	purple moor-grass	6-8	—	S
Other plants	*Agrostis tenuis*	common bent-grass	6-8	U—	R
	Anthoxanthum odoratum	sweet vernal-grass	4-6	U	S
	Carex spp.	sedges		U	S
	Deschampsia flexuosa	wavy hair-grass	6-7	—	XS
	Erica cinerea	bell-heather	7-9	—	XW
	Eriophorum angustifolium	common cotton-grass	5-6	—	R
	Galium saxatile	heath bedstraw	6-8		S
	Holcus lanatus	Yorkshire fog	6-9	U	S
	Juncus articulatus	jointed rush	6-9	U	S
	Juncus conglomeratus	conglomerate rush	5-7		S
	Juncus squarrosus	heath rush	6-7	—	S
	Myrica gale	bog myrtle, sweet gale	4-5		W
	Nardus stricta	mat-grass	6-8		XR
	Narthecium ossifragum	bog asphodel	7-9	—	R
	Scirpus caespitosus (= *Tricophorum caespitosum*)	deer-grass	5-6		S

[1] Months of flowering [2] Distribution

Cotton-grass moor

This occurs in northern Britain, in regions that have high rainfall and a cool climate. Such regions occur in parts of the Pennines, and in the Scottish uplands. The rate of decay is slow, so that a thick layer of peat is built up, creating boggy conditions. The dominant plant is *Eriophorum vaginatum* (cotton-grass), which is easily recognized by its bulbous inflorescences with their silvery glumes. It is *not* a grass, but a member of the Cyperaceae, the sedge family. In common with many members of this family it can live in waterlogged, acid soils, low in dissolved oxygen. In these conditions few other species can flourish, so that large areas of cotton-grass moor are covered with nothing but *Eriophorum*. As mentioned above, this plant may also be found dominating the more waterlogged parts of molinia-grass moors.

The deep, damp, cold peat, in which oxygen and mineral ions are in short supply, presents very severe conditions for plants, but where the lie of the land is suitable there may be streams that bring about an improvement of the conditions. These streams will cut deep channels in the soft peat, and where there are many streams and their branches, there will be relatively raised islands of peat between them. Water will tend to drain more easily from these raised 'peat hags', giving drier conditions more suited to plant growth, so allowing species other than *Eriophorum* to become established.

FLORA OF COTTON-GRASS MOOR

(for abbreviations, see endpapers)

DESCRIPTION	SCIENTIFIC NAME	COMMON NAME	M[1]	D[2]	HABIT
Dominant	*Eriophorum angustifolium*	common cotton-grass	5-6	—	R
On hags and better-drained areas	*Calluna vulgaris*	heather ling	7-9	—	XW
	Empetrum nigrum	crowberry	5-6	N—	W
	Erica cinerea	bell-heather	7-9	—	XW
	Erica tetralix	cross-leaved heath	5-6	—	XW
	Molinia caerulea	purple moor-grass	6-8	—	S
	Nardus stricta	mat-grass	6-8		XR
	Rubus chamaemorus	cloudberry			WR
	Rubus fruticosus	blackberry, bramble	6-9	U	W
	Vaccinium myrtillus	bilberry, blaeberry, whortleberry, huckleberry	4-6	—	XW

[1] Months of flowering [2] Distribution

Nardus-grass moor

This is found in drier conditions than either of the two previous moorland types, mainly in northern and western Britain. The dominant plant is *Nardus stricta* (mat-grass). It requires less damp and less acid soils than the other two moorland 'grasses'. *Nardus* may be recognized by its spiky leaves with their blades rolled up to form cylinders (a xeromophic habit), and by its inflorescences, which are simple in structure consisting of two rows of spikelets arranged along *one* side of the main axis. As suggested by its name, *Nardus* forms a dense mat of interwoven individuals, usually excluding most other vegetation. The continuous cover of *Nardus* is broken only occasionally by patches of *Deschampsia flexuosa* or by other species where there are local variations in condition due to streams or trodden paths.

The flora lists on pp. 118-122 have not included any ground-layer plants. Where thick tussocks or dense mats of the moor-grasses dominate there is no scope for any vegetation beneath. However, there are always places such as rocky outcrops, walls, and stream banks where mosses such as the bog mosses (*Sphagnum*) can become established. The damp conditions make moorland suitable for the growth of liverworts, such as *Marchantia polymorpha*, *Gymnocolea inflata*, and *Diplophyllum albicans*.

Calluna heath

This is found on dry, sandy soils in lowland areas of England, such as parts of the New Forest, parts of Ashdown Forest, and in Dorset, Surrey and Cornwall. There are also areas of calluna heath in East Scotland. Since the soil is sandy, with plenty of gravel, and since it is thin and drains easily, the few mineral salts it contains are quickly leached out by rainfall, leaving a soil very deficient in mineral ions. The exposed situation of the soil, the thinness of the soil, and the ease with which it drains all combine to produce a very dry soil. The growth of plants is sparse, so little humus is produced and this rots slowly owing to the dry conditions. This thin surface layer of peat absorbs the rainfall, so that little penetrates to the soil below. Later, when the sun shines, the surface layer is warmed, and the water quickly evaporates, leaving the peat as dry as before, and ensuring that little water gets down to the roots of plants in the soil beneath.

Under these conditions *Calluna* becomes the dominant plant. Like many other plants growing on moors and heaths it possesses xeromorphic features (p. 116). On heathland, xeromorphism is easy to explain, for there the plants experience drought frequently and for extended periods. In moorland areas this seldom happens, for soil water is usually in excess. The moorland plants experience acid soil conditions, the oxygen content of the soils is low, and the soils are cold. All these factors contribute towards reducing the ability of the plants to absorb water. In addition, the shoots of the plants are exposed to strong winds, and occasionally to intense sunshine. because of the exposed, treeless sites. Under these circumstances the plants will tend to lose water at a greater rate than they can absorb it. *In effect*, they will suffer from drought—not *physical* drought, as on dry heathland, but *physiological* drought. Against either form of drought xeromorphic features are a valuable safeguard.

The dominant *Calluna* is a relatively tall plant and the shade it casts may exclude other plants over large areas of heathland. Only a few mosses and lichens are found beneath the *Calluna*. Liverworts are absent, for they grow well only in damp places.

The low mineral level of the soil due to leaching has already been mentioned. This level is further reduced because the leaves and stems of *Calluna* do not rot readily, especially under the dry conditions of heathland. As on moorland, we find many species that are able partly to overcome the shortage of mineral elements by obtaining the most important one, nitrogen, from the atmosphere. We find *Calluna* and *Erica* spp., which have mycorrhizas, many leguminous plants (*Ulex*, *Sarothamnus*, *Lotus*) and, in the wetter areas, some insectivorous plants.

The mention of wetter areas is a reminder that even in a generally dry area such as a heath there may be regions where, owing to the configuration of the land, water accumulates and causes the formation of bogs. Here one finds vegetation more akin to that of moorland.

As on moorland, the vegetation is constantly being removed either by fire or by grazing animals. Both these factors operate toward keeping the vegetation unchanged, with the heath plants dominant. If sheep and, in particular, rabbits were to be excluded and fires prevented, then seeds of trees would enter these areas and seedling trees would become established. These would grow and eventually shade the *Calluna* or other heath and moor dominants, leading to the establishment of woodland. This is not to say that heaths are treeless, for there are always combinations of factors that in restricted areas allow a few trees to reach the size beyond which they are safe from grazers and have overtopped the dominant *Calluna*. In other areas factors may swing in favour of *Ulex* or some others of the woody heathland plants. *Ulex* may be locally dominant over large areas. The most usual species of *Ulex* found on heaths is *Ulex europaeus*, but other species may be found growing with it. *Ulex gallii* is often found, particularly in western Britain, and *Ulex minor* is common in heathland in the south.

FLORA OF CALLUNA HEATH

(for abbreviations, see endpapers)

DESCRIPTION	SCIENTIFIC NAME	COMMON NAME	M[1]	D[2]	HABIT
Dominant	*Calluna vulgaris*	heather, ling	7-9	–	XW
On dry heath	*Agrostis canina*	brown bent-grass	6-7		XS/R
	Deschampsia flexuosa	wavy hair-grass	6-7	–	XS
	Erica cinerea	bell heather	7-9	–	XW
	Galium saxatile	heath bedstraw	6-8		S
	Larix decidua	larch	4-5		W
	Linum catharticum	purging flax	6-9	+	a
	Lotus corniculatus	bird's-foot trefoil, eggs-and-bacon	6-9	U	S
	Nardus stricta	mat-grass	6-8		XR
	Polygala serpyllifolia	thyme-leaved milkwort	3-8		S
	Pteridium aquilinum	bracken	7-8	U–	XR
	Rubus fruticosus	blackberry, bramble	6-9	U	W
	Sarothamnus scoparius	broom	5-6		XW
	Ulex europaeus	furze, gorse, whin	3-6	U	W
	Vaccinium myrtillus	bilberry, blaeberry, whortleberry, huckleberry	4-6	–	XW
	Viola palustris	marsh violet	4-7		R
Lichen	*Cladonia* spp.			U	LL
Mosses	*Ceratodon purpureus*	purple-fruiting heath moss			CM
	Dicranum scoparium	lesser fork moss			FM
	Hypnum cupressiforme	cypress-leaved feather-moss			FM
	Pleurozium schreberi	red-stemmed feather-moss			FM
	Polytrichum spp.	hair-mosses			CM
In bogs	*Carex* spp.	sedges		U	S
	Drosera rotundifolia	sundew	6-8	–	VS
	Erica tetralix	cross-leaved heath	5-6	–	W
	Eriophorum angustifolium	common cotton-grass	5-6	–	R

[1] Months of flowering [2] Distribution

DESCRIPTION	SCIENTIFIC NAME	COMMON NAME	M	D	HABIT
	Juncus spp.	rushes			S
	Myrica gale	bog myrtle, sweet gale	4-5		W
	Pinguicula vulgaris	common butterwort	5-7		V Rootless bud
	Salix repens	creeping willow	4-5		W
	Scirpus caespitosus	deer-grass	5-6		S
Mosses in bogs	*Polytrichum commune*	common hair-moss			CM
	Sphagnum compactum	bog moss			CM
	Sphagnum tenellum	bog moss			CM
Liverworts in bogs	*Aplozia crenulata*				LLW
	Calypogeia trichomanis				LLW

Animals of heath and moor

For small ground-living invertebrate animals, heaths and moors provide plenty of microhabitats. We may find here most of the species encountered in other inland habitats, though in pools and streams there are fewer species than is usual in freshwater habitats, presumably because of the acid reaction of the peaty water. Ground beetles, springtails, centipedes, millipedes, and many microscopic soil invertebrates are common. Earthworms are less common, for they do better in soils containing calcium. This may be well illustrated on an old tennis lawn; over the years the repeated marking of the court with chalk will have enriched the soil immediately below the lines, so that in later years, even when the grass is *un*marked, the design of the court can still be made out because of the large number of worm casts formed in calcium-rich soil. It is not easy to explain why earthworms should not flourish in calcium-deficient soils, but it is clear that snails, which require large quantities of calcium for the manufacture of their shells, will be relatively uncommon on heaths and moors. Woodlice, common in most damp habitats, are often found on moors, but less commonly on heaths where the dry soils and the exposure to wind and sunshine create conditions they cannot tolerate. The wind is a limiting factor for flying insects. The larger insects, which are able to control their flight path in the stronger winds, are at an advantage, so bumble bees and other large insects are the chief pollinators of *Calluna* and other insect-pollinated heath plants. Moths are common on moors and heaths. Usually their wings bear a fine-grained brown-white pattern which camouflages them against their normal background of *Calluna* or *Erica* on which they rest during the day. An example is the common heath moth (*Ematurea atomaria*). Spiders are plentiful, building their webs on the vegetation. In heathland, ground-living jumping and wolf spiders are common. These do not build webs, but catch their prey directly, either by jumping on to them or by chasing them.

Of the vertebrate animals, amphibia are rare on heaths because of their thin skins, which do not allow them to live in dry places. Reptiles such as lizards and snakes can often be found. Birds find few suitable nesting places owing to the lack of trees and large shrubs. A few, such as the skylark, nest on the ground. The majority of other birds found on heaths and moors will be visitors from nearby woods and hedgerows, or coastal birds.

Mammals are plentiful on moors and heaths, provided they are able to obtain shelter from the cold. Sheep are protected against all but the coldest weather by their thick coat. They exert an important effect on the vegetation of large areas by grazing, and help to prevent the moorland vegetation from progressing to scrub or woodland. Rabbits are common on heathland, and gain protection by burrowing. They are preyed on by stoats and weasels.

GRASSLAND

British grasslands are of four main types:

Siliceous grassland
Neutral grassland
Chalk grassland
Limestone grassland

Siliceous grassland

This occurs on older rocks, such as are found in the Pennines, the Lake District, and in Wales. Here the soil is low in calcium and is thin. Since these are regions of high rainfall the meagre supply of calcium is leached out and the soil reaction becomes acid. The dominant plant is *Agrostis tenuis* (common bent-grass), but other grasses may be present, such as *Festuca ovina* and *Molinia caerulea* in wetter areas, and *Nardus stricta* in drier areas. In wetter areas sedges and rushes are often found replacing the grasses. Siliceous grassland is generally used for sheep grazing, and it is probable that the grazing prevents most other species of plants from becoming established; the list of plants commonly found is certainly a short one. It is also noticeable that the majority of plants in the list are calcifuge species (—), as might be expected on calcium-deficient soils. If ungrazed, siliceous grassland shows succession to molinia moor (p. 120) or to calluna moor (p. 116).

FLORA OF SILICEOUS GRASSLAND

(for abbreviations, see endpapers)

DESCRIPTION	SCIENTIFIC NAME	COMMON NAME	M[1]	D[2]	HABIT
Dominant	*Agrostis tenuis*	common bent-grass	6-8	U–	R
Other plants	*Calluna vulgaris*	heather, ling	7-9	–	XW
	Carex spp.	sedges		U	S
	Deschampsia flexuosa	wavy hair-grass	6-7	–	XS
	Empetrum nigrum	crowberry	5-6	N–	W
	Erica tetralix	cross-leaved heath	5-6	–	XW
	Festuca ovina	sheep's fescue	5-8	U	S
	Holcus mollis	creeping soft-grass	6-7		S
	Juncus spp.	rushes		U	S
	Nardus stricta	mat-grass	6-8		XR
	Narthecium ossifragum	bog asphodel	7-9	–	R
	Molinia caerulea	purple moor-grass	6-8	–	S
	Pteridium aquilinum	bracken	7-8	U–	XR

[1] Months of flowering [2] Distribution

Neutral grassland

This is found on the relatively fertile clays and loams of central and south-east England. It has arisen on areas that were previously forest and which have been cleared and taken under cultivation. The soil is neutral or perhaps slightly acid, so that the growth of many species is possible. If neutral grassland were left to follow its natural succession, forest would probably regenerate. This is prevented by the use to which the grassland is put. It is used either as 'pasture' grazing land for sheep, cattle, and horses, or as 'meadow' when the grass is allowed to grow and is mown once or twice a year to yield hay for feeding to farm stock. The kind of vegetation found in a pasture or meadow will depend on the treatment it has previously received: cutting, herbicides, insecticides, watering, shading, the nature of manuring, the amount of grazing, the kinds of animals that have grazed there, whether it is harrowed or rolled, how often it is mown, and at what seasons. The interaction of the activities of meadow or pasture plants, the farm animals and the farmer is complex, so that only the main points can be dealt with here.

The dominant plants are those grasses that are palatable to farm animals. They have a fine growth and interweave to form a grassy mat. The commonest meadow grasses are listed below. Some of these are commonly found growing in places other than meadows or pasture, but often grasses found in meadows and pasture will belong to imported strains, such as Italian rye-grass which is grown to provide an early spring food supply. Many special varieties of these meadow grass species have been bred, and mixtures of seeds of these varieties with seeds of specially bred varieties of leguminous plants are sown to provide the best food for farm animals. Some farmers will plough their land and sow these seed mixtures to produce a meadow or pasture that is allowed to grow for a few years. Such a meadow or pasture is called a 'ley' and is part of a crop rotation, being eventually ploughed up and the land used for some other farm crop such as wheat. The ley functions to feed livestock and to improve soil fertility. By the ploughing in of organic material, the texture and nutrient level of the soil is increased. By the action of the nitrogen-fixing bacteria of the root nodules of the leguminous plants, atmospheric nitrogen is utilized, and the overall nitrogen content of the soil is thus increased when the remains of these plants are ploughed in.

The composition of the vegetation of a recently sown ley will depend very much on what seed mixture was used. For example, a seed mixture useful for providing plenty of hay and capable of supporting occasional grazing is:

Grasses: Perennial rye-grass, variety S.24
Italian rye-grass, variety S.22
Danish cocksfoot
Cocksfoot, variety S.37
Timothy, variety S.51

Legumes: Red clover, English broad-leaved
Red clover, English late-flowering
Red clover, variety S.123
White clover, variety S.100

Such a mixture would contain about one third of perennial ryegrass, which is regarded as one of the best ley grasses. Many other mixtures are sown, depending on situation and requirements, and many would contain fewer seed types, but in general leys are sown with an average of two grasses and two legumes. Commonly, these are varieties of the species marked * on p. 132, with the occasional addition of varieties of lucerne and sainfoin, both leguminous plants.

If leys are left unploughed for several years their composition will gradually change. Some species may disappear and new species will appear as weeds. Many of these are deep-rooted plants with a basal rosette of leaves. These escape competition with the shallow-rooted grasses, are not affected by ploughing, and quickly recover from treatment with herbicides. Among the weeds we find various buttercups, ragwort, bracken, plantains, thistles, docks, and some of the weeds of arable land (List N, p.93). Some of these are not eaten by grazing animals: for example, bracken, ragwort, the buttercups, and Yorkshire fog are distasteful to animals, and the thistles are clearly unpalatable.

If the ley is used for pasture rather than as a meadow, these plants will in time increase by growth, while the more palatable ley grasses will be kept in check by grazing. Such plants can often be seen on old pasture, standing up untouched above the closely grazed turf. As time goes on, the vegetation becomes less and less like that of the original ley and eventually becomes permanent grassland; that is, land that is no longer part of a farm rotation, but for various reasons is kept permanently under grass. The term 'permanent' is, of course, only relative. Permanent grassland can at any time be ploughed to make way for other farm crops or for a new ley.

In a long-term ley or in a permanent grassland there will have been sufficient time for new species to become established under the influence of diverse environmental factors. Then it will be possible to investigate the variations in the vegetation from one part of the pasture or meadow to another. Certain parts of a pasture may be less frequented by farm animals or, in a meadow, the shading caused by adjacent hedges or trees may have caused local variations in light intensity that have affected the vegetation. Streams and ponds can introduce further variation. Fields in low-lying areas beside rivers or canals are frequently put down to permanent grassland, for they are less suited as arable land. They are permanently damp and often flooded. Meadow grasses are less well adapted for such conditions and their place is gradually taken by typical waterside plants (p. 53). Among the commonest are *Cardamine pratensis*, *Filipendula ulmaria*, *Caltha palustris*, *Ranunculus repens*, and *Deschampsia caespitosa*. *Deschampsia* is distinguishable from the meadow grasses by its tussock-forming habit (see Plate 10). In the wettest areas there may be clumps of rushes (*Juncus* spp.) or of sedges (*Carex* spp.).

Animal life in meadows and pastures is varied. The range of ecological niches is large and the environment is favourable so that there are plenty of small invertebrates, both in the soil and associated with the plants. There are also many birds, feeding either on the small invertebrates or on the seeds of the grasses and legumes. A common animal of meadows and pastures is *Talpa europaea*, the mole, which feeds mainly on earthworms, and raises its conspicuous chains of mole-hills over a surprisingly large area. The prolific insect life of the soil and plant layer makes this a good habitat for the mole, and also for other insect-eating mammals such as the hedgehog, and the shrews. Where the vegetation grows taller and is not grazed or mown, conditions are milder at soil level, and among such vegetation are found the tunnels and runs of the short-tailed vole (*Microtus*) and the bank vole (*Clethrionomys*).

FLORA OF NEUTRAL GRASSLAND

(for abbreviations, see endpapers)

DESCRIPTION	SCIENTIFIC NAME	COMMON NAME	M[1]	D[2]	HABIT
Meadow grasses	**Dactylis glomerata*	cock's-foot	5-7	U	SZ
	**Festuca pratensis*	meadow fescue	5-8		SZ
	**Lolium multiflorum* (var. *italicum*)	Italian rye-grass	5-8		RZ
	**Lolium perenne*	perennial rye-grass	5-8		RZ
	**Phleum pratense*	Timothy	7	U	SZ
Other plants	*Caltha palustris*	kingcup, marsh marigold	3-7	U	R
	Cardamine pratensis	lady's-smock, cuckoo flower	4-6	U	S
	Carex spp.	sedges		U	S
	Cirsium arvense	creeping thistle	7-9	U	Ro
	Cirsium vulgare	spear thistle	7-10	U	b
	Deschampsia caespitosa	tufted hair-grass	6-8	U	S
	Filipendula ulmaria	meadow-sweet	6-9	U	R
	Holcus lanatus	Yorkshire fog	6-9	U	S
	Juncus spp.	rushes		U	S
	Lathyrus pratensis	meadow vetchling	5-8	U	
	Lotus corniculatus	bird's-foot trefoil, eggs-and-bacon	6-9	U	S
	**Medicago lupulina*	black medick	4-8	U	a
	Plantago lanceolata	ribwort	4-8	U	S
	Plantago major	great plantain	5-9	U	S
	Pteridium aquilinum	bracken	7-8	U–	R
	Ranunculus auricomus	goldilocks	4-5		S
	Ranunculus bulbosus	bulbous buttercup	5-6	U	C
	Ranunculus ficaria	lesser celandine	3-5	U	T
	Ranunculus repens	creeping buttercup	5-8	U	S
	Senecio jacobaea	ragwort	6-10	U	b/S
	Trifolium dubium	lesser yellow trefoil	5-10	U	a
	**Trifolium pratense*	red clover	5-9	U	SZ
	**Trifolium repens*	white clover	6-9	U	SZ
	Vicia spp.	vetches		U	S

[1] Months of flowering [2] Distribution * See p. 130

Chalk grassland

This is found on the South Downs, the Isle of Wight, and several other areas of southern England. The shallow, alkaline soil is exposed to wind and sunshine, giving rise to dry soil which is well aerated. The soils, being derived from the chalky bed-rock, are rich in calcium, and a high proportion of the species growing there are calcicolous (+). The two dominant plants are the grasses *Festuca ovina* (sheep's fescue) and *Festuca rubra* (red fescue), which make a fine turf, very suitable for grazing. The continual grazing by sheep and rabbits keeps the vegetation low, and many other species—provided they are adapted to the dry exposed habitat—are able to establish themselves among the dominant grass plants. The grazing reduces the competition for light. Plants that have a basal rosette of leaves are well suited to these conditions. The leaves are close to the soil, so that transpiration is minimal and they escape the grazing animals. Several plants with the rosette habit (J) are found in the list on p. 134. Plants with a creeping or prostrate habit (p) are also for the most part able to escape the effects of grazing.

Although the soil of these regions is characteristically rich in calcium, local variations can occur. For example, rain will leach the calcium (and other minerals) from the surface layers of soil, so that shallow-rooted plants such as the grasses will be growing in soil largely deficient in calcium. None of the grasses listed are calcicoles. By contrast, the deeper-rooted plants such as *Poterium* and *Helianthemum* root in moist, calcium-rich soil. On the tops of the downs, where the land is level and rain does not run off, the most extreme leaching occurs. The soil may be calcium-deficient to such a depth that a deep-rooting calci*fuge* such as *Calluna vulgaris* can become established. This is the plant most often associated with moorland and heathland (pp. 116-126), and in becoming established on the tops of downs it gives rise to yet another type of heathland—*chalk-heath*. Later, this area may be invaded by the smaller trees and bushes such as *Sorbus aria*, *Viburnum lantana*, and *Crataegus monogyna*. These provide support for climbing plants and shade beneath which other species become established, leading to the formation of *chalk scrub*. Chalk scrub is also found in old chalk pits and on the edges of beechwoods. Later, other trees such as beech, yew, or box may grow up within the scrub. These will eventually grow to full size, and chalk scrub will give way to a woodland community.

FLORA OF CHALK GRASSLAND

(for abbreviation, see endpapers)

DESCRIPTION	SCIENTIFIC NAME	COMMON NAME	M[1]	D[2]	HABIT
Dominant	*Festuca ovina*	sheep's fescue	5-8	U	S
	Festuca rubra	red fescue	5-7	U	S
Other plants	*Achillea millefolium*	yarrow, milfoil	6-8	U	Sp
	Anacamptis pyramidalis (and other rare orchids)	pyramidal orchid	6-8	S	p Root tubers
	Asperula cynanchica	squinancy wort	6-7	+	Sp
	Briza media	quaking grass, doddering dillies	6-7		S
	Carduus nutans	musk thistle	5-8	S+	bJ
	Carlina vulgaris	carline thistle	7-10	S+	bJ
	Centaurea nigra	lesser knapweed, hardheads	6-9	U	SJ
	Cirsium acaulon	stemless thistle	7-9	S+	TJ
	Daucus carota	wild carrot	7-8		b
	Helianthemum chamaecistus	common rockrose	6-9	+	Wp
	Hieracium pilosella	mouse-ear hawkweed		U	Rp
	Lotus corniculatus	bird's-foot trefoil, eggs-and-bacon	6-9	U	S
	Ononis spinosa	restharrow	6-9		Wp
	Phleum pratense	Timothy	7	U	S
	Plantago media	hoary plantain	5-8	S+	SJ
	Poterium sanguisorba	salad burnet	5-8	S+	S
	Primula veris	cowslip	4-5	S	RJ
	Scabiosa columbaria	small scabious	7-8	S+	TJ
	Taraxacum officinale	common dandelion	3-6	U	TJ
	Thymus drucei	wild thyme	5-8		Wp

[1] Months of flowering [2] Distribution

FLORA OF CHALK SCRUB

(for abbreviations, see endpapers)

DESCRIPTION	SCIENTIFIC NAME	COMMON NAME	M[1]	D[2]	HABIT
Trees and shrubs	*Crataegus monogyna*	hawthorn	5-6	U	W
	Juniperus communis	juniper	5-6		W
	Ligustrum vulgare	privet	6-7		W
	Rubus fruticosus	blackberry, bramble	6-9	U	W
	Sambucus nigra	elder	6-7	U	W
	Sorbus aria	white beam	5-6	S	W
	Thelycrania sanguinea	dogwood	6-7	S	W
	Viburnum lantana	wayfaring tree, mealy guelder rose	5-6	S+	W
Climbers and scramblers	*Clematis vitalba*	traveller's joy, old man's beard	7-8	S+	W
	Hedera helix	ivy	9-11	U	W
	Rosa canina	dog rose	6-7		W
	Solanum dulcamara	bittersweet, woody nightshade	6-9		W
Other plants	*Brachypodium sylvaticum*	slender false-brome	7		S
	Mercurialis perennis	dog's mercury	2-4	U	R
	Pastinaca sativa	wild parsnip	7-8	S	b

and plants of chalk grassland.

[1] Months of flowering [2] Distribution

Limestone grassland

This is found in the Pennines, in Wales, Westmorland and in the Cotswold area. The soil conditions are similar in many ways to those of chalk grassland. The dominant grass is *Festuca ovina*, which makes a good grazing turf. The soil is dry, for limestone is very permeable to water, and it is alkaline and rich in calcium. Thus, as in chalk grassland, many species are calcicolous (+), and there are several with rosette habit (J) or scrambling habit (p).

FLORA OF LIMESTONE GRASSLAND

(for abbreviations, see endpapers)

DESCRIPTION	SCIENTIFIC NAME	COMMON NAME	M[1]	D[2]	HABIT
Dominant	*Festuca ovina*	sheep's fescue	5-8	U	S
Other plants	*Asperula cynanchia*	squinancy wort	6-7	+	Sp
	Brachypodium pinnatum	tor-grass, heath false brome	7	+	S
	Campanula glomerata	clustered bell-flower	5-9	+	S
	Helianthemum chamaecistus	common rockrose	6-9	+	Wp
	Helicotrichon pratensis	meadow oat	6	+	S
	Helicotrichon pubescens	hairy oat	6-7	+	S
	Koeleria cristata	crested hair-grass	6-7	+	S
	Luzula campestris	field woodrush	3-6	U	S
	Plantago media	hoary plantain	5-8	S+	SJ
	Poterium sanguisorba	salad burnet	5-8	S+	S

and many plants of chalk grassland.

[1] Months of flowering [2] Distribution

COASTAL HABITATS

Environmental factors

The coast is a boundary zone between the purely terrestrial environment of inland areas and the utterly different marine environment of the oceans. The margins of woods and the banks of streams are also boundary zones, and in these regions we find many species that are not found in either of the habitats separated by the boundary zone. Many of these species show special adaptations to the particular conditions existing in the boundary zone and are less suited to living on either side of it. It is noticeable that a very large proportion of the plants in the flora lists (pp. 162-173) have what is termed *coastal distribution* (C), being common on coasts but rare inland.

The assemblages of organisms in boundary zones do not form communities in the sense that those in a wood or in a pond form an almost self-contained group of closely interacting species. Boundary zones are frequently too narrow and too discontinuous to allow this. Plants on the margin of a wood may compete for root space with the trees within the wood, and plants growing at the edge of a pond may be a source of food for animals from the adjacent meadow yet, at the same time, their dead leaves may fall into the pond and, through the food network, provide food for the animal life of the pond. In the same way, many seashore animals derive their food from plants and animals of the open sea, and may, in turn, be fed on by animals from inland areas.

The coastal habitats have some features in common with other boundary zones, but here the similarities end. The boundary zones on the edges of woods or ponds are regions of transition between the habitats on either side of the zone. There are gradual or abrupt changes in the environmental factors as one passes from one habitat, across the boundary zone, to the other habitat. For example, as one passes from the interior of a wood to an open meadow there is a gradual increase in mean light intensity and perhaps a decrease in mean relative humidity. Usually the conditions in the boundary zone are *intermediate* between those found in the areas on either side. This is not so in coastal habitats. There the conditions are in the main quite unlike conditions found either inland or at sea. For example, the powerful action of waves beating on the shore is not experienced by the inhabitants of

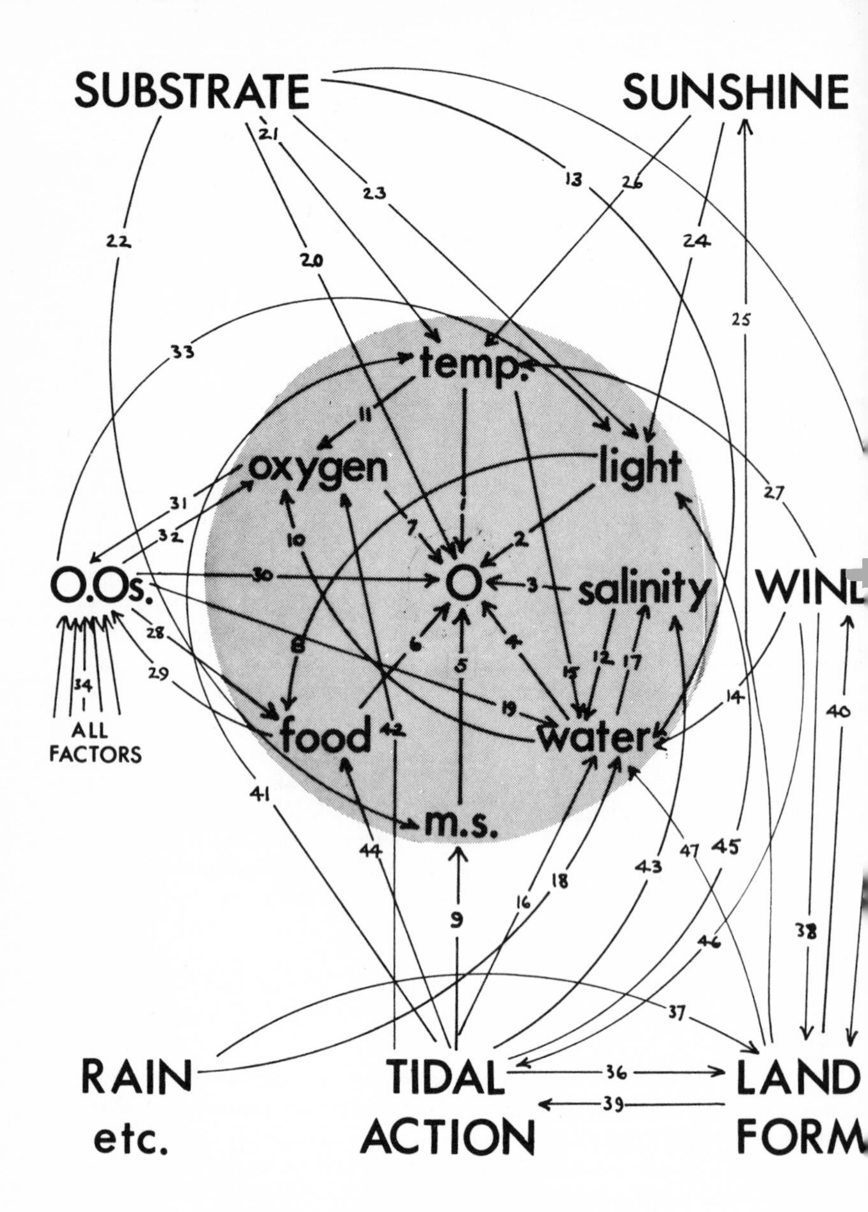

Fig. 3. The interactions of factors on the seashore.

O (in centre)	=	organism
temp.	=	temperature
m.s.	=	mechanical shock
O.Os	=	other organisms
rain etc	=	rain, dew, springs, streams and other sources of fresh water

Grey area represents the immediate environment.

inland areas or by dwellers in the open sea. Another point of difference is that the coastal zone comprises several distinct zones, each with its own combination of environmental factors and its own complement of characteristic species.

The main environmental factors of coastal habitats are shown in Fig. 3, where their many interactions are indicated by connecting arrows. In the discussion that follows, the numbers in parentheses refer to the numbered arrows of the diagram.

In the centre of the diagram is an organism, and surrounding it, within the grey circle, are the seven most important factors of its immediate environment:

1. *Temperature*. This affects the rate of metabolism of plants and of poikilothermic (= cold-blooded) animals. Apart from birds and a few mammals living in or on sand dunes and cliffs, all coastal animals are poikilothermic, so are greatly influenced by temperature. In addition, they are killed by extremely high or low temperature.

2. *Light*. This is of great importance to plants, for photosynthesis, but of less direct importance to animals. If light intensity is low, the growth of plants is restricted and there is less food for herbivorous animals (8 and 6) and thus for carnivores.

3. *Salinity*. This affects the osmotic relationships between the organism and its surroundings. Sea water provides a fairly *constant* salinity and a pH of 8·2 to 8·4, which is well suited to the animals that live there. Variations in salinity may cause difficulty.

4. *Water*. An essential constituent of living organisms, and essential for many of their activities, such as feeding, respiration, locomotion, and reproduction.

5. *Mechanical shock*. A factor of special importance in tidal regions, where wave action exerts enormous forces (9). Many shore creatures are adapted to minimize the worst effects of mechanical shock.

6. *Food*. The sea provides a plentiful supply of plankton, which is the food of many animals of the tidal zone. The plants of this zone are bathed by the sea water, which is rich in mineral ions. By contrast, on rocky cliffs and on sand dunes food material for plants and animals is scanty, and populations are correspondingly sparse.

7. *Oxygen*. This is essential for the respiration of plants and animals. Carbon dioxide, too, is essential, for it is the raw material of photosynthesis, and where oxygen is mentioned in the discussions, this additional factor applicable to plants must also be taken into account. Most marine organisms obtain their oxygen (and carbon dioxide) in solution in the sea water, so the availability of oxygen depends on whether a sufficient quantity of

sea water is present (10). The solubility of gases is lower at higher temperatures, so the availability of oxygen will be less when the water is warm (11).

The factors listed above are those by which an organism is most directly affected. An organism suffers if there is water shortage (4), but it is not significant to the organism if this shortage is the result of excessive salinity causing osmotic interference with water uptake (12), or rapid drainage because of the porosity of the substrate (13), or excessive evaporation due to strong wind (14) or high temperature (15), or the withdrawal of water by the ebbing tide (16). Whichever one or more of these factors might be responsible, the organism is short of water, and is affected accordingly. Shortage of water can have an indirect effect on the organism by increasing the salinity (17 and 3). Equally serious effects can occur when salinity is suddenly reduced by rainfall or by water from a freshwater stream that crosses the coastal zone (18). At low tide the organisms living among seaweed or thickly encrusting growths of certain animals may escape the worst effects of desiccation (19). So do those that live in tubes or burrows in a sandy or muddy substrate (13), and those that burrow in rocks or live in crevices between rocks (13).

Thus, the seven factors of the immediate environment have direct effects on the organism and interact with each other. The level at which each of these factors operates is determined by other environmental factors, which we might term *determining factors*. These, together with the seven immediate factors, make up the total environment by which the organism is affected. The determining factors are shown at the periphery of Fig. 3. Of these the substrate could possibly be considered to be part of the immediate environment, for many organisms are profoundly influenced by the substrate (20). Burrowing and boring animals are obvious examples, and seaweeds will not survive unless they become attached at an early stage to a firm, rocky or wooden substrate. At the opposite extreme, a shifting substrate such as a sand dune or a shingly beach is a difficult place on which to become established. So substrate is of direct importance to many organisms, but it is included among the determining factors because it interacts with several of them, whereas the immediate factors do not. Apart from its direct effects on organisms, the substrate has many indirect effects on the immediate environment, such as:

(*a*) its water-absorbing capacity, and the ease with which it drains (13);
(*b*) its specific heat and its colour, which determine its rate of heating and cooling (21);
(*c*) the degree to which it provides a medium into which animals can burrow, or provides crevices into which animals can retreat, to obtain protection from the mechanical shock of waves (22);
(*d*) the amount of turbidity it produces if it becomes suspended in water, with the consequent effect on light intensity below the water surface (23).

Whereas substrate is a constant feature of a given locality, sunshine, and hence light intensity, varies from hour to hour, from day to night, from day to day, and from season to season (24). In general, coastal habitats are devoid of trees, so that there is full exposure to the sun's radiation, except where cliffs or high dunes cast shadows (25). During periods of sunshine, temperature may rise rapidly (26). Conversely, in dull weather or at night there are rapid falls. The result is that coastal organisms are subject to wide and rapid changes of temperature. The range will be most extreme in areas that are not covered by tides (see p. 144). Wind is another climatic condition that affects local temperature (27). Coasts usually experience stronger winds than occur inland, so wind is an important factor for coastal organisms. Wind also influences water supply, for it increases the rate of evaporation by removing humid air (14).

No organism lives in complete isolation from others, and there are complex interactions between individuals of the same or of different species (see p. 158) Some of these interactions are shown on Fig. 3:

(*a*) one organism may feed on another (28 and 6);
(*b*) one organism may compete with another for food (29 and 6);
(*c*) one organism may be fed upon or grazed upon by another (30);
(*d*) animals may compete for supplies of oxygen (31 and 7)—or for carbon dioxide in the case of plants;
(*e*) photosynthesizing plants provide oxygen for use by animals (32 and 7);
(*f*) large seaweeds shade smaller ones (33);
(*g*) large seaweeds provide a moist microhabitat for small animals during periods of low tide (19);
(*h*) like the organisms in the centre of the diagram, the other organisms are affected (perhaps in differing ways) by their own immediate environment and by the determining factors (34) in a manner too complex to be shown in the diagram and extremely difficult to disentangle in the field.

Another determining factor is land form—the presence or absence of cliffs, bays, banks, beaches, dunes, spits, and the like, taking into account their form and location. Land form is determined largely by geological processes, involving the nature of the substrate (35), the erosive or depository action of the sea (36), of streams and rivers (37), and of wind (38). While these forces are shaping and re-shaping the land and sea bed, the contours that have thus been generated have their effects on tides (39) and wind direction (40). These effects operate not only on the large scale but also on the small scale. On one side of a small boulder the action of wind and tide may be very different from their action on the other.

The coasts are regions of continuous change. The pattern of life on the coasts is determined by events of the past as well as by those of the present, and is altering slowly, day by day, as a result of the constant but gradually changing action of wind and waves.

The most characteristic coastal factor, tidal action, has been left until last. Cliffs and dunes may be beyond the range of tidal activity, but on shore the effects of tides are of paramount importance; they are also very complex, being superimposed on all the interactions that have been noted so far.

In principle, the tides advance and recede approximately twice a day, under the influence of the gravitational attraction of the moon. At any given spot within the tidal zone of the shore an organism is subjected to the following cycle of events:

A. *At high tide*

The organism is immersed, and its immediate environment can be summarized as:

Temperature—(41) sea temperature, which is uniform, changing little with the seasons; warmer than land temperature in winter; cooler than land temperature in summer.

Water—plentiful supply (16).

Wave action—negligible.

Oxygen—plentiful supply (42), also plenty of carbon dioxide for plants.

Salinity—normal sea water, of constant composition suited to marine organisms (43).

Food—plentiful supply of marine life, especially plankton, organic debris, mineral ions (44).

Light—perhaps slightly reduced in intensity (45), due to suspended mineral particles, etc. An important change is spectral composition. Light of shorter wavelengths is absorbed to a greater extent than light of longer wavelengths so that as depth increases, light becomes bluer. It is said that the orange-red pigment (fucoxanthin) of the brown algae is an adaptation towards making more efficient use of the bluer light. The main effect of light is upon plants, which are able to grow down to a depth of about 50 m.

Locomotion is easy.

Danger of predators (30) is high.

In general, conditions are FAVOURABLE.

B. *Tide is receding*

The organism is at the water's edge:

Temperature—about sea temperature, as above.

Water—plentiful.

Wave action—the pounding and rapid currents cause severe mechanical strain.

Oxygen—plentiful supply: wave action gives full aeration of sea water.
Salinity—about the same as normal sea water.
Food—about the same as before, but feeding difficult.
Light—about full daylight.
Locomotion—impossible.
In general, conditions are UNFAVOURABLE.

C. *At low tide*

The organism is exposed to:

Temperature—very high if sunny; very low at night, in winter, or if wind is strong.
Water—gradually drains away, and may evaporate from surface layers. A few centimetres below sand, sufficient water may remain until the tide returns. Rock pools retain a good supply of water. Supply may be replenished by fresh water (rain, dew or streams).
Wave action—nil, though there may be salt spray.
Oxygen—lack of water and increasing temperature will reduce the availability of dissolved oxygen.
Salinity—increasing as water evaporates from pools; decreasing during rain or by irrigation from streams. So may range widely and rapidly.
Food—feeding impossible for most shore animals, except in pools.
Light—full daylight.
Locomotion—difficult or impossible.
In general, conditions are UNFAVOURABLE.

D. *Tide is advancing*

The organism is at the water's edge:

Temperature—falling, though at first the sea in the tidal zone may have been warmed by passage over hot rocks or sand, so the fall may be delayed.
Other conditions as for B, which in general are UNFAVOURABLE.

During this cycle there is only one stage at which the conditions are suited to the normal activities of life of shore organisms.

Now consider some of the ways in which this cycle may be modified. Frequently the cycle occurs once daily instead of twice, or one tide of the day may be higher than the other. Under these circumstances an organism on the upper part of the shore will experience only one high tide, and will spend the larger part of the day under low-tide conditions. These are unfavourable and become even less favourable the longer the period between one high tide and the next. On the lower shore, organisms may experience only one low tide daily. The heights of tides vary from day to day, so that each month there are two periods of spring tides, which give the highest high tides and the lowest low tides, alternating with two periods of neap tides, when the waters advance and recede by the least amount. Organisms above the level of the highest neap tides are immersed once or twice daily during the period of spring tides but may be exposed for a week or more during neap tides. Conversely, on the lower shore organisms are submerged continuously except during the lowest spring tides. This regime is further modified, for at certain times of the year the sequences of spring tides (of which there are two each month) are alternately large and less large in range. Organisms high on the shore are submerged for only a few hours each day for a few days, and then remain exposed for as long as three weeks until the next sequence of highest spring tides occurs.

As a result of the variations in tides there is a gradation in the amount of immersion and exposure, from the highest level reached by the highest spring tides down to the lowest level to which they recede. High on the shore there are periods of days or weeks without immersion and low on the shore long periods without exposure. Only in the mid-shore region is there the regular twice-daily routine of exposure and immersion.

This variable tidal behaviour may be further modified by local conditions, the most important effects being:

(*a*) *Land form* (39). If coasts are steep, the full impact of the waves is rapidly broken and wave action is very violent. Conversely, on gently sloping shores the force of the waves is dissipated over a wide area and wave action is slight. This difference exists not only between two major regions of the coast but also between smaller sub-regions. On a boulder the waves may break with much greater force than on the sloping beach alongside, with differing effects on the plant and animal populations in the two situations. In places of violent action much spray is generated and organisms above the tide level will be well saturated with sea water. This *in effect* raises the tide level, so that marine animals and plants are found higher up the shore than usual. On a gently sloping, smooth shore the breaking waves may flood some distance up the shore; this too raises the effective tide level.

(*b*) *Wave force*. This depends on the degree to which the locality is exposed to open sea. In an exposed site, wave action is stronger, raising the effective tide level when waves spray or flood after striking the shore. The direction of the prevailing wind exerts an additional influence here (46).

(*c*) *Drainage*. When the tide has receded, the rate of drainage is dependent on the slope of the shore (47), its smoothness, and the type of material (13) of which it is composed. Slow drainage extends phase A (p. 142) until after phase B. Rapid drainage brings about a rapid transition from phase B to phase C. The effects of rapid drainage are enhanced in sunny (26 and 15) and windy (14) sites.

(*d*) *Cover by seaweeds*. These retain water between their fronds, creating a moist microhabitat in which small animals can remain active during low tide. This shortens phase D for these animals. A dense cover of seaweed also dampens the force of wave action, and to a lesser extent so does a dense cover of barnacles or mussels.

Fig. 3 shows that tidal action has direct effects on *all* the factors of the immediate environment (9, 16, 41, 42, 43, 44, 45). The effect of such a varying factor as tidal action, impinging on the complex network of interactions already described, makes it a formidable undertaking to attempt to analyse the living conditions of a given region of coast. It is often impossible to disentangle all the factors involved. It is also very difficult to predict what will be the effects of a change in any one of them. In spite of this, it is possible to undertake worth-while studies in such areas, and the wide variety of life found on the coasts, with its manifold adaptations to the extremes of conditions found there, is a powerful incentive for intensive studies. One can observe broad regions of zonation on the shore and investigate to what extent these are due to tidal zones, and discover if this pattern is modified by local conditions. If there are two adjacent areas, differing in only one or two factors—such as substrate, aspect, or slope—the problem of analysis is simplified and comparisons can profitably be made. On dunes and cliffs the situation is less complex because of the absence of the overriding effect of tides. From such studies some general principles emerge and these will be outlined below. However, when working in any given area one must be prepared to find wide departures from the patterns given here, for local conditions can cause significant variations on the standard patterns. Working out which local conditions are responsible for such departures is another interesting task.

The seashore—general

Two major determining factors on the seashore are the substrate and the degree of exposure to wave action. In general, we classify shores as rocky, sandy, shingly, or muddy. Muddy shores are not found in regions exposed to strong wave action, for the small mud particles are quickly dispersed by water currents to settle in more sheltered regions. Conversely, in sheltered regions we seldom find rocky shores, for sand and mud settle there, covering the rocks. The nature of the substrate has such a large effect on the seashore life that we shall consider each type of shore separately.

Rocky shores

These contain many different kinds of microhabitat so that there is a far wider variety of plant and animal life here than on sandy, shingly, or muddy shores. The firm substrate is less affected by erosion or deposition than sand, shingle, or mud, and it provides a firm and permanent place of attachment for plants and animals. In time, well defined zones of plant and animal life can develop on this stable substrate and the effects of tide level and of other environmental factors can be clearly seen. The main zones found on a rocky shore are best defined by reference to the chief organisms they contain. Lewis (see bibliography) distinguishes the following zones (Fig. 4):

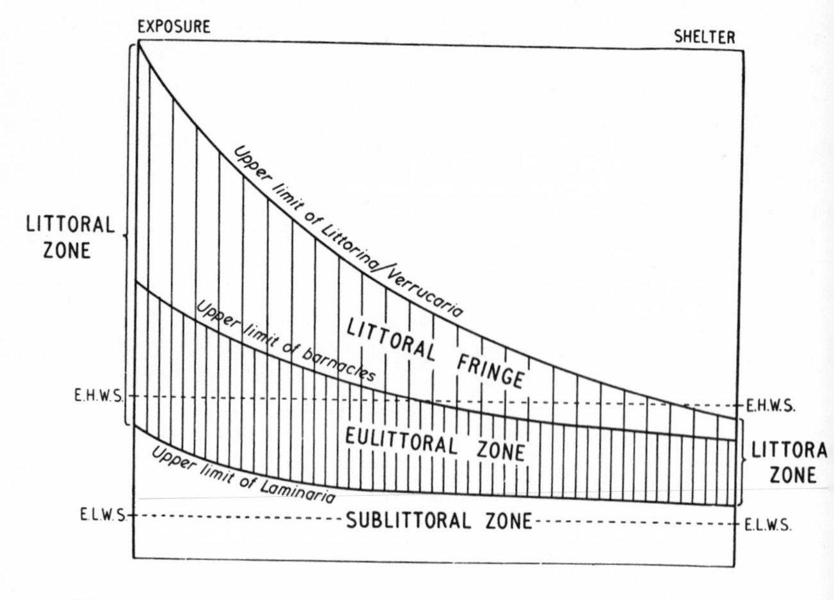

Fig. 4. Zonation on rocky shores (from LEWIS, see Book List p. 177, no. 27).
EHWS = extreme high water of spring tides
ELWS = extreme low water of spring tides

1. *Littoral fringe:* dominated by littorinids (periwinkles) and the black lichen, *Verrucaria.* This lichen covers the rocks with a thick, continuous film, so that the rock may appear to be coloured black. This zone is the highest shore zone to contain true marine species. Above it we find terrestrial species such as the orange lichens *Caloplaca* and *Xanthoria.* On the more sheltered coasts the amount of *Verrucaria* is reduced and the orange lichens partly occupy the upper region of the littoral fringe. Occasionally specimens of the red alga, *Pelvetia,* are found in the lower part of this zone. Other organisms to be found in the littoral fringe are listed on p. 152 (List C).

2. *Eulittoral zone:* dominated by acorn barnacles (*Balanus, Chthamalus*) and limpets (*Patella*). *Chthamalus* generally occupies a zone above *Balanus.* This is the intertidal zone, lying approximately between the highest and lowest of spring tides. But the location of the zone with respect to tide level varies according to the exposure of the shore (Fig. 4) and effective tide levels may be raised or lowered by local conditions (p. 144). In general, the more exposed shores exhibit zonation of marine species further above the highest spring tide level. The eulittoral zone is occupied by many species other than the dominant ones; these are listed on p. 152 (List D).

3. *Sublittoral zone:* dominated by the brown alga, *Laminaria*, or related species. Organisms in this zone are covered by the sea for long periods, or continuously. Conditions are generally favourable, so this zone supports a large number of organisms of a wide range of species (List E, p. 154).

It is seldom that all those species listed will be found in one locality. The assemblage of species found varies from area to area, and in some places one may find a locally dominant species that is rare elsewhere and so is not listed in this book. On steep, rocky faces, where the distance between high and low tide levels may be no more than two metres, there will not be room for zones of each species of the larger algae to become established, so only one or two species will be present.

In the lists on pp. 151-157 the organisms have been classified under their evolutionary groups. Some of the characteristics of these groups are given below:

Animals

Protozoans. These single-celled animals are plentiful, the commonest being the foraminifera, which secrete around themselves a protective shell of calcareous material. Owing to difficulties of identification and the fact that there are so many types of greater interest, protozoans have not been included in the lists.

Poriferans (the sponges). They are located mainly in the sublittoral zone, though a few are found in the eulittoral zone. These animals are not well adapted to stand long periods of exposure. They need a firm substrate for attachment, which is provided by the rocks. They feed on microscopic organisms (including protozoans), extracting them from the current of sea water which passes through their body, and which they generate by the lashing of flagella borne on special cells.

Coelenterates. This group includes the sea anemones, the jellyfish, the hydroids, and the corals. True corals are rare on the coasts of Britain, and none are included in these lists. *Alcyonium*, p. 154, is not a true coral. The sea anemones and hydroids need a firm substrate for attachment, and this is provided by the rocks or the fronds of larger seaweeds. Being more or less fixed they cannot retreat down-shore as the tide recedes, so they need to be able to resist desiccation at low tide. Anemones withdraw their tentacles into the mouth, and the gelatinous body, contracted into an almost hemispherical shape, exposes the minimum of body surface to the air at low tide. The polyps of the hydroids can retract into their skeletal covering at low tide. The jellyfish are mobile and, therefore, are found mainly in the sea. Some are found stranded in pools on rock flats or on sandy patches.

Flatworms (*Platyhelminthes*). These are probably quite common, but not easily seen, so they do not often appear among fauna lists of the sea-shore. At low tide they may be found in damp places under stones.

Nemertine worms (*Nemertini*). The ribbon worms belong to a group with relatively few species, several of which can be found on the shore. They are usually found under stones.

Annelid worms. Almost all annelid worms found on the seashore belong to the polychaete group. The oligochaete group, the group to which the earthworms belong, is not represented. The polychaetes are divided into two main groups:

(1) *Errant polychaetes.* These swim or crawl freely. They are found in pools, under stones, and buried in sand or gravel.

(2) *Sedentary polychaetes.* These live in burrows or tubes, from which they project their feeding apparatus, or through which they generate a feeding current and strain off microscopic food material as it is wafted through. Some species build calcareous tubes on the rocks or on the fronds of the larger seaweeds. Others form tubes of a mixture of mucus and sand or mud. The tubes can be found in sand or mud, in crevices between rocks, of between the holdfasts of the larger seaweeds.

The polychaetes are soft-bodied, thin-skinned animals, so they are not able to resist desiccation. The sedentary forms gain valuable protection from predators and from desiccation by their burrows or tubes. Polychaetes are absent from the littoral fringe, and are abundant only in the sublittoral zone.

Arthropods. The only group of arthropods well represented on the seashore is the crustacean group. These occur in the eulittoral and sublittoral zones where at low tide they occur under stones, in crevices, or in pools. Among the commonest are the acorn barnacles (*Balanus*, *Chthamalus*) which are sessile and are constructed very differently from other crustaceans. Insects are the major group of arthropods, for they account for about three-quarters of all known animal species, yet this group is very poorly represented on the seashore. The larvae and adults of the midge, *Clunio marinus*, may sometimes be found among *Laminaria* in the sublittoral zone. In the littoral fringe one may find species of bristle-tails, a primitive insect group. Apart from these true shore-dwellers there are often visitors such as flies and beetles feeding in heaps of rotting seaweed on the strand line. Under stones in the eulittoral fringe and upper shore a few members of the other arthropod groups, the spiders and the centipedes, may be found.

Molluscs. Apart from the cephalopods (squids, octopuses, cuttlefish) which are sea creatures and are found on shore only when stranded, the molluscs have a hard shell or shells. The gastropods have a single shell. This serves as a protection against both desiccation and wave action. Thus protected, they are well able to withstand the rigours of seashore life, and are found in great numbers and great variety in all zones. The several species of the genus *Littorina* (periwinkles) may be the dominant or only genus present over large expanses of shore. The animal withdraws into its shell at low tide, and can survive long periods of desiccation. Often the periwinkles congregate in rock crevices, even in quite shallow depressions, thus gaining additional protection from the waves. Should they become dislodged, the almost spherical shell rolls freely in the currents. A sphere has great mechanical strength (so has a dome, such as the human skull) and resists impact and crushing. When covered with water, *Littorina* and other gastropod molluscs feed by rasping. Many graze on algae, but some such as the whelks are carnivorous. The limpets show a different kind of adaptation to wave action. Their conical shell fits tightly against the rock surface and is held firmly in place by the powerful muscles of the limpet's body. The animal cannot be dislodged by wave action, and the close-fitting conical shape ensures that water currents flow smoothly past, meeting the least possible resistance.

The bivalve molluscs (those with two shells, hinged together) are more sedentary in habit than the gastropods. They feed by taking in a current of water and filtering microscopic food particles from it. They do not need to move far and can feed even when buried in sand, with only the feeding openings projecting above the surface. Many resist desiccation and wave action by burying themselves in sand. Bivalve molluscs, whether they bury themselves or not, are secure against desiccation and predators inside the two shells which fit exactly together and are kept firmly shut by means of strong muscles within. Some bivalves secrete a mass of threads (a byssus) which anchors the shell to gravel, stones or rocks. The mussel (*Mytilus*) does not bury itself but has a very strong byssus, and remains firmly attached in spite of wave action.

A few of the bivalve molluscs have their shells shaped so as to give a cutting edge. These molluscs can bore holes in rock or wood, and so gain excellent protection.

Bryozoans. These grow in colonies on the surface of seaweeds or sometimes on rocks. Each small animal is enclosed in its own protective case of a hard material, into which it can withdraw when necessary.

Echinoderms. This group includes the starfish and sea urchins. A few specimens may be found stranded at higher levels, but these animals are found chiefly in the sublittoral zone. They are animals more at home in the deeper coastal water and do not seem to be adapted to survive the rigours of the tidal zone.

Chordate animals include—

Tunicates. These are sedentary animals, attached to rocks or seaweed. Each animal has a tough outer body layer, the tunic. The animal feeds by extracting materials from a current of water that it passes in through its mouth and out by another opening. Tunicates are very common in the sublittoral zone.

Fish. Being highly mobile these are generally able to keep pace with the advancing or receding tides, so are usually found at the water's edge or farther out. A few, such as gobies, remain in rock pools and are usually hidden in crevices between the rocks.

Birds. These have not been included in the lists on pp. 151-157 because they are not true shore animals. The shore is no place for an object as fragile as a bird's egg, and the young are helpless when first hatched. Fledgling and adult birds of many species may frequently be seen in all regions of rocky and sandy shores, and on the sea itself, searching for food. Gulls of all kinds come there, turning over materials on the strand line or hunting among the rocky pools. A list of birds commonly found on coasts is on p. 174.

Mammals. The only marine mammals likely to be found on rocky shores are seals, such as *Halichoerus grypus*, the grey seal, and *Phocas vitulina*, the common seal. Seals are not encountered often, except in special localities where they may congregate in great numbers and breed. Terrestrial mammals may come on to the shore to feed when the tide is out. The commonest such mammals are *Rattus rattus*, the black rat, *Rattus norvegicus*, the brown rat, *Sorex araneus*, the common shrew, *Sorex minutus*, the pygmy shrew, and *Homo sapiens*. The latter may be seen collecting fish from the water's edge, and hunting for shrimps, prawns, various molluscs, and some seaweeds such as dulse, laver, and carragheen moss. By his collecting activities on the seashore, as well as by his recreational activities, he can exert a major influence on the plant and animal life of the shore.

Plants

Algae. The three main groups of algae are all well represented on the seashore. These are the Chlorophyceae (the green algae), the Phaeophyceae (the brown algae) and the Rhodophyceae (the red algae). Algae are predominantly aquatic plants and those of the seashore are well adapted to such an environment. The algal body is flexible yet resistant to tension. It is able to bend with the currents yet does not break away from the holdfast by which it grips the rocks. The tissues of algae are very mucilaginous and therefore less liable to desiccation. Algae have the ability to recover after prolonged periods of severe desiccation, conditions under which most herbaceous land plants wilt and die very quickly. The slippery, mucilaginous substance with which the algae are covered renders them less likely to tangle and tear when agitated by waves.

Lichens. These are not truly marine organisms, but a few are able to resist occasional immersion and the action of salt spray. They live in or just above the littoral fringe. One of them, *Verrucaria*, is extremely successful there, and is usually the dominant organism of that zone.

Flowering plants. The only one to appear in the tidal region is *Zostera*. Other species may occasionally appear on the shore, especially if it is adjacent to a salt marsh (pp. 165-167) or if a freshwater stream crosses the shore. In common with the ferns and gymnosperms (also absent from shores), they rely on establishing a root system for obtaining water and mineral nutrients. This is not possible in the rocky or sandy substrates of shores, and the high salinity makes the absorption of water and mineral ions extremely difficult. Only salt-marsh plants are adapted to absorb from saline solutions (p. 165).

FAUNA AND FLORA OF ROCKY SHORES

Abbreviations

Status	D	=	usually the dominant species in this zone
	E	=	characteristic of exposed shores
	½E	=	characteristic of semi-exposed shores
	S	=	characteristic of sheltered shores
Habit	A	=	attached to large algae
	B	=	boring in rock
	P	=	usually in pools
	T	=	in tubes built on rocks
	Ta	=	in tubes built on large algae
	Trc	=	in tubes in rock crevices
	U	=	usually under stones

GROUP	SCIENTIFIC NAME	COMMON NAME	STATUS	HABIT
A. Immediately above highest tidal level (organisms inhabiting lower levels, but commonly found stranded)				
Coelenterates	*Aurelia aurita*	common jelly fish		
	Pleurobrachia pileus	sea gooseberry		
Molluscs	*Buccinium undulatum*	common whelk		egg cases
	Sepia officinalis	cuttlefish		skeleton
and the empty shells of many species.				
Fish	*Raja* spp.	rays and skates		egg cases, (mermaid's purses)
	Scyliorhinus spp.	dogfish		egg cases, (mermaid's purses)
B. Living on rocks in or just above the splash zone				
Lichens	*Caloplaca* spp.		S	
	Lecanora atra			
	Lichina confinis		E	
	Ramalina spp.			
	Xanthoria parietina		S	

GROUP	SCIENTIFIC NAME	COMMON NAME	STATUS	HABIT

C. Littoral fringe

GROUP	SCIENTIFIC NAME	COMMON NAME	STATUS	HABIT
Flatworms	*Procerodes ulvae*			U
Arthropods (Crustaceans)	*Armadillidium vulgare*	pillbug		U
	Leander spp.	prawn		P
	Ligia oceanica	sea slater		U
	Praunus spp.	chameleon shrimp		P
	Talitrus spp.	sand hopper		among seaweed
Molluscs	*Lasaea rubra*	(bivalve)		
	Littorina neritoides	small periwinkle	D	
	Littorina saxatilis	rough periwinkle		
Lichens	*Lichina confinis*		E	
	Verrucaria maura		D	
	Xanthoria parietina		S	
Algae—				
(Brown)	*Pelvetia canaliculata*	channel wrack	½E/S	
(Red)	*Porphyra* spp.	laver	E	

D. Eulittoral zone

GROUP	SCIENTIFIC NAME	COMMON NAME	STATUS	HABIT
Poriferans	*Leucoselenia* spp.			P
	Hymeniacidon sanguinea			
Coelenterates	*Actinea equina*	beadlet anemone		
	Anemonia sulcata	snakelocks anemone		
	Dynamena (= *Sertularia*) *pumila*	(hydroid)		A
	Laomedia (= *Campanularia*) *flexuosa*	(hydroid)		P
	Obelia spp.	(hydroid)		A
	Tealia felina	dahlia anemone		P
Flatworms	*Procerodes ulvae*	(flatworm)		U
Annelid worms	*Eulalia viridis*	(errant polychaete)		U

GROUP	SCIENTIFIC NAME	COMMON NAME	STATUS	HABIT
	Sabellaria	(sedentary polychaete)	½E	Tr
	Spirorbis spp.	(sedentary polychaete)		Tr/Ta
Arthropods (Crustaceans)	*Balanus* spp.	acorn barnacle	D	
	Carcinus maenas	shore crab		
	Chthamalus stellatus	acorn barnacle	D	
	Gammarus spp.	sand hopper		U
	Leander spp.	prawn		P
	Porcellana spp.	porcelain crab		U
	Praunus spp.	chameleon shrimp		P
	Sphaeroma rugicauda	(isopod)		
Molluscs	*Aeolida papillosa*	common grey sea slug		
	Gibbula cineraria	grey topshell		
	Lasaea rubra	(bivalve)		
	Lepidochitona cinereus	coat-of-mail shell (chiton)		
	Littorina littoralis	flat periwinkle		
	Littorina littorea	edible periwinkle		
	Littorina saxatilis	rough periwinkle		
	Mytilis spp.	mussel		
	Nucella (= *Thais*) *lapillus*	dog whelk		
	Patella spp.	limpet		
Bryozoans	*Bugula turbinata*	sea mat		
	Flustrella hispida	sea mat		A
	Membranipora membranaceae	sea mat		A
Lichens	*Lichina confinis*			
	Lichina pygmaea			
Algae— (Green)	*Cladophora* spp.			
	Enteromorpha spp.			
	Ulva lactuca	sea lettuce		P
(Brown)	*Ascophyllum nodosum*	egg wrack	½E/S	

GROUP	SCIENTIFIC NAME	COMMON NAME	STATUS	HABIT
	Ectocarpus spp.			
	Fucus serratus	serrated wrack	½E/S	
	Fucus spiralis	spiral wrack	½E/S	
	Fucus vesiculosis	bladder wrack	½E/S	
	Laminaria saccharina	kelp		P
	Mesogloia vermiculata			
	Pelvetia caniculata	channel wrack	½E/S	
	Sphacelaria spp.			A
(Red)	*Catenella repens*			
	Ceramium spp.			P
	Chondrus crispus	carragheen moss		
	Corallina officinalis	coral weed		P
	Gigartina stellata	carragheen moss		
	Laurencia pinnatifida			
	Membranoptera alata			PA
	Plumaria elegans			
	Porphyra spp.	laver	E	
	Rhodymenia spp.	(including dulse)		A

E. Sublittoral zone

GROUP	SCIENTIFIC NAME	COMMON NAME	STATUS	HABIT
Poriferans	*Grantia compressa*			
	Halichondria panicea	bread-crumb sponge		
	Hymeniacidon sanguinea			
	Sycon coronatum			
Coelenterates	*Actinia equina*	beadlet anemone		
	Adamsia palliata	(anemone on shell of hermit crab)		
	Alcyonium digitatum	deadman's fingers, soft coral		
	Anemone sulcata	snakelocks anemone		
	Aurelia aurita	common jellyfish		
	Dynamena (= *Sertularia*) *pumila*			A

GROUP	SCIENTIFIC NAME	COMMON NAME	STATUS	HABIT
	Kirchenpaueria pinnata			P
	Laomedia (= *Campanularia*) *flexuosa*			P
	Metridium senile	plumose anemone		
	Obelia spp.	(hydroid)		A
	Pleurobrachia pileus	(sea gooseberry)		P
	Sargartia elegans	(anemone)		
	Tealia felina	dahlia anemone		P
Nemertine worms	*Nemertopsis flavida*	(ribbon worm)		PU
Annelid worms	*Eulalia viridis*	(errant polychaete)		U
	Nereis spp.	ragworm		
	Polydora spp.	(sedentary polychaete)		Trc
	Pomatoceros triqueter	(sedentary polychaete)	S	Tr
	Sabellaria spp.	(sedentary polychaete)	S	Tr
	Serpula vermicularis	(sedentary polychaete)		Tr
	Spirorbis spp.	(sedentary polychaete)		Tr/Ta
Arthropods—(Crustaceans)	*Carcinus maenas*	shore crab		
	Grangon vulgaius	shrimp		
	Eupagurus bernhardus	common hermit crab		P
	Gammarus spp.	sand hopper		U
	Idotea spp.	(isopod)		P
	Leander spp.	prawn		P
	Porcellana spp.	porcelain crab		U
	Praunus spp.	chameleon shrimp		P
Molluscs	*Archidoris pseudoargus*	sea lemon		
	Gibbula cineraria	grey topshell		
	Hiatella arctica	(bivalve)		B
	Lacuna vincta	banded chink shell		

GROUP	SCIENTIFIC NAME	COMMON NAME	STATUS	HABIT
	Littorina littoralis	flat periwinkle		
	Littorina littorea	edible periwinkle		
	Mytilus spp.	mussel	S	
	Pholas dactylus	common piddock		B
	Venerupis pullastra	pullet carpet shell		
Bryozoans	*Bugula turbinata*	(sea mat)		
	Flustrella hispida	(sea mat)		A
	Membranipora membranaceae	(sea mat)		A
Echinoderms	*Asterias rubens*	common starfish		
	Echinus esculentus	common sea urchin		
	Ophiothrix fragilis	(starfish)		
Tunicates	*Ascidia* spp.	(sea squirt)	S. & W. coasts	
	Botryllis schlosseri	star sea squirt		
Fish	*Gobius* spp.	goby		P
Algae—				
(Green)	*Enteromorpha* spp.			
	Ulva lactuca	sea lettuce		P
(Brown)	*Alaria esculenta*	kelp	DE	
	Chorda filum	bootlace		
	Ectocarpus spp.			
	Himanthalia elongata	thong weed		
	Laminaria digitata	kelp	D½E	
	Laminaria hyperborea	kelp	D½E	
	Laminaria saccharina	kelp	DS	P
	Mesogloia vermiculata			
	Sacchoriza polyschides	kelp		
(Red)	*Gigartina stellata*	carragheen moss		
	Lomentaria articulata			
	Membranoptera alata			PA

GROUP	SCIENTIFIC NAME	COMMON NAME	STATUS	HABIT
	Plumaria elegans			
	Porphyra spp.	laver	E	
	Rhodymenia spp.	(including dulse)		A

F. The sea edge

GROUP	SCIENTIFIC NAME	COMMON NAME	STATUS	HABIT
Coelenterates	*Aurelia aurita*	common jellyfish		
Arthropods (Crustaceans)	*Leander* spp.	prawn		
	Porcellana spp.	porcelain crab		
	Praunus spp.	chameleon shrimp		
Fish	*Anguilla anguilla*	common eel		
	Blennius spp.	blenny		
	Centronotus gunnelus	butterfish		
	Cottus scorpius	father lasher		
	Gobius spp.	goby		
	Onos spp.	bearded rockling		
	Spinachia vulgaris	15-spined stickleback		

Interactions between the organisms of the rocky shore

With so many different species occupying so many microhabitats it is no wonder that the pattern of interaction between species is complex. Fig. 5 shows the principal forms of interaction. These interactions are superimposed upon the interactions between environmental factors (Fig. 3) and the result is the intricate boundary community of the seashore. This community is probably never stable, for at any time one or more of the organisms involved can be subjected to a severe depletion of numbers—perhaps by a completely chance event such as a succession of exceptionally high tides, or a season exceptionally favourable for the multiplication of one of its predators. A variation in one population will have a chain effect. An excess of limpets may lead to a reduction in mussel larvae because the limpets prevent the larvae from settling. Later, this will lead to a reduced population of adult mussels, which will give opportunity for an increase in the population of barnacles owing to a relaxation in the competition for space. A chance effect acting on *one* species can in time affect *all* species. Life on the shore passes through a succession of such changes, each spreading over several seasons. There probably never is a stable climax of fauna and flora, but instead a series of patterns that *appear* stable but which, over a long period, are seen to follow on from one to the other. The unravelling of these patterns and of the interactions that produce them is one of the fascinating aspects of studying shore life.

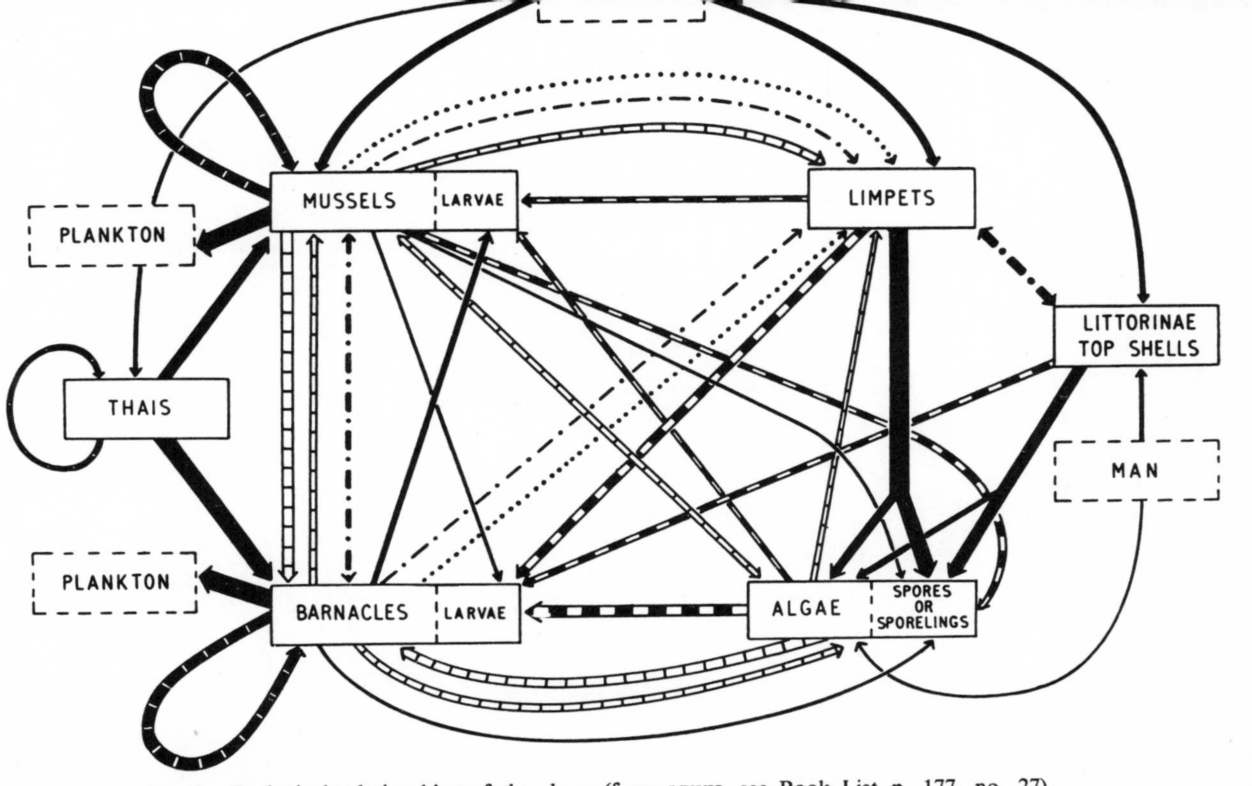

Fig. 5. Ecological relationships of the shore (from LEWIS, see Book List p. 177, no. 27).

Sandy shores

There is less to say under this heading, partly because the general points have already been covered in the section on rocky shores, and partly because the sandy shore offers fewer microhabitats. Consequently, fewer species are found there.

On rocky shores many plants and animals, including all the dominant ones, rely on their firm attachment to the rocky substrate. A sandy substrate provides no such anchorage, and this alone makes it impossible for poriferans, sea anemones, most hydroids and corals, many gastropod molluscs, bryozoans, tunicates, algae, and lichens to inhabit sandy shores.

On exposed coasts particularly, the sand shifts with the tides and so makes the development of well defined zones of organisms impossible. This is why location has been related to tide levels in the lists. The shore locations are defined here as:

Upper shore — above the mean high tide level.
Middle shore— between mean high tide and mean low tide.
Lower shore — below mean low tide.

The violent action of the waves on exposed shores produces large-scale shifting of the sand, and throws it up in steeply sloping banks. These drain quickly when the tide goes out, and few animals find favourable conditions there. On sheltered coasts the sand lies in gently sloping beaches, and there is less shifting. On such wide sand flats the animal population is larger. The animals gain protection from tidal action by spending the low-tide period, at least, beneath the sand. Some bury themselves in the sand, others live in burrows or build tubes in the sand. At low tide, a few centimetres below the surface of apparently dry sand, sufficient water remains between the sand particles to provide a humid environment and allows these sand-dwellers to survive. Another refuge on a sandy beach is the occasional small boulder or pebble. Beneath these may be found small animals such as some of the smaller of the crustaceans, and attached to these stones there may be small specimens of algae such as *Laminaria saccharina*. Another place where animals may be found at low tide is beneath piles of seaweed and driftwood on the strand line.

The pattern of life on a sandy shore is much simpler than that on a rocky shore, yet there is still plenty to be found and investigated. If a sandy shore has a rocky area adjacent to it, some interesting comparisons can be made.

FLORA AND FAUNA OF SANDY SHORES

Abbreviations (see also endpapers)

Zonation	L	=	lower shore
	M	=	middle shore
	S	=	in sea
	U	=	upper shore
Habit	Bm	=	burrows in mud
	Bs	=	burrows in sand
	Is	=	buries itself in sand
	M	=	on muddy shores, as well as on sand
	P	=	in pools
	Tm	=	forms tube in mud
	Ts	=	forms tube in sand

DESCRIPTION	SCIENTIFIC NAME	COMMON NAME	M[1]	Z[2]	D[3]	HABIT
Coelenterates	*Adamsia palliata*	(anemone)		L (on hermit crab shells)		
	Hydractinea echinata	(hydroid)		L		
Annelids	*Arenicola* spp.	lugworm		L		Bs/Bm
	Glycera spp.	(errant polychaete)		L		M
	Lanice conchilega	(sedentary polychaete)		L		Ts
	Nephthys spp.	including catworm (errant polychaetes)		ML		M
	Nereis spp.	ragworm		ML		Is
	Sabella pavonina	(sedentary polychaete)		L		Ts/Tm
Arthropods (Crustaceans)	*Crangon vulgaris*	shrimp		L		P
	Eupagurus bernhardus	hermit crab		L		

[1] Months of flowering [2] Zonation [3] Distribution

DESCRIPTION	SCIENTIFIC NAME	COMMON NAME	M	Z	D	HABIT
	Orchestia spp.	sandhopper		U		under weeds
Molluscs	*Cardium edule*	cockle		ML		
	Dosinia spp.	artemis		L		Is
	Ensis spp.	razor shell		L		
	Spisula spp.	trough shell		L		Is
	Tellina spp.	tellin		ML		Is
	Venus spp.	venus		L		Is
Echinoderms	*Astropecten irregularis*	(starfish)		S		Is
	Echinocardium cordatum	heart urchin		ML		Is
	Ophiura spp.	brittle star		L		Is
Fish	*Ammodytes* spp.	sand eels		SL		Is
	Trachinus spp.	weever fish		L		Is
Algae—(Brown)	*Laminaria saccharina*	kelp		ML		on stones
Flowering plants	*Eryngium maritimum*	sea holly	7-8	U	C	S
	Honkenya peploides		5-8	U	C	S
	Polygonum aviculare	knot-grass	7-10	U	U	S
	Salicornia spp.	glasswort	8-9	U	C	aY
	Salsola kali	saltwort	7-9	U	C	a
	Zostera spp.	eel-grass		ML	C	M

On the strand-line there may be species of list A, p. 151.
In the sea there may be species of list F, p. 157.

Muddy shores

These are found only in the most sheltered areas, often close to the mouths of estuaries, where there is little water movement. Such shores have many of the characteristics of sandy shores but slope imperceptibly. The mud forms a compact and relatively stable substrate. Many animals of sandy shores can be found here, especially those normally living on the surface. Fewer burrowing and tube-building animals occur, since the compactness of the mud makes burrowing difficult. There is also plenty of organic material in the mud, with the result that oxygen supply is deficient a few centimetres below the surface. This does not affect tube-dwellers, which circulate water through their tubes, but animals that bury themselves will suffer. When the area is covered by water, the fine mud particles suspended in the water tend to block the filter-feeding mechanisms of some species, including many tube-dwellers, and this is a further disadvantage of life on a muddy shore. Salt marshes often form on the landward side of muddy shores (pp. 165-167).

Shingly shores

Shingle consists of large pebbles, and forms a loose substrate easily disturbed by wave action. The extreme mobility of shingle makes it very unsuitable for colonization by plants and animals, except perhaps on the landward side of a ridge or spit. Materials thrown over the ridge by exceptionally high seas can accumulate on the landward side and there decay, adding humus to the shingle. This improves its water-retaining properties, which are initially very poor. Rain and dew provide water, and the shingle becomes colonized with plants (list p. 164). Since such a region may be frequently drenched with salt spray, and occasionally flooded by a high tide, conditions will be slightly saline and plants suited to salty conditions will be common. Two examples of such salt-tolerating, or halophytic plants, are *Silene* and *Halimione*. Succulent plants such as *Honkenya*, *Sedum* and *Erygium* are also well adapted to life on shingle.

Thus, under the right conditions, it is possible for the landward side of a ridge of shingle to become colonized, and animal life will move in. On the crest of the ridge, exposure to wind will usually prevent plants from growing. On the seaward side shingle is usually devoid of life.

FLORA OF SHINGLY SHORES

(for abbreviations, see endpapers)

SCIENTIFIC NAME	COMMON NAME	M[1]	D[2]	HABIT
Cakile maritima	sea rocket	6-8	C	a
Calystegia soldanella	sea bindweed	6-8	C	S
Erygium maritimus	sea holly	7-8	C	S
Glaucium flavum	yellow horned poppy	6-9	C	b/T
Halimione portulacoides	sea purslane	7-9	C	RWY
Honkenya peploides		5-8	C	S
Polygonum aviculare	knotgrass	7-10	U	S
Rumex crispus	curled dock	6-10	U	T
Sedum acre	wall-pepper, biting stonecrop	6-7		S
Silene maritima	sea campion	6-8	C	SY

[1] Months of flowering [2] Distribution

Salt marsh

On the banks of wide estuaries, the rate of flow of river water is low. At high tide, water from the sea may flood up the estuary, reversing the current twice a day and causing periods of stagnation. Under such conditions fine soil particles suspended in the river water settle down and mud banks will be formed. Vegetation develops on these mud flats, and there is a succession very similar to that from fresh water to fen (pp. 60-61). The waterlogged soil is low in oxygen content, and its mobility makes rooting difficult, so that few species are able to colonize the area at first. Most important is the influence of the sea, for the water is sometimes fresh, and sometimes almost as saline as sea water, and these phases alternate according to the tides. The mixing of fresh water with sea water produces brackish water, with salinity intermediate between that of the sea and the river.

Plants adapted to live in conditions of excessive salinity are called halophytes. Their cell sap has an osmotic pressure higher than that found in other plants, so that they are better able to absorb water from saline solutions. Many have succulent leaves and stems which contain water storage tissues. Water is absorbed rapidly when salinity is low (at low tide or after heavy rain) and stored for use during periods when salinity is high. The succulent habit is characteristic of xeromorphs, too, for these, like the halophytes, live in conditions in which water is difficult to obtain. Salt-marsh plants show other adaptations, such as aeration tissues in their roots, to offset the effects of oxygen deficiency in the mud, and long rhizomes and roots, which help them to become anchored in the mobile substrate.

An examination of a salt-marsh soon shows that there is zonation. The zones are not necessarily clear-cut, and the zonation may vary with local conditions. The most usual type of zonation is outlined below. It illustrates the succession that occurs during the colonization of the mud flats.

The chief zones are:

1. *Low salt marsh.* This is nearest the water's edge, and is covered, twice daily by the high tides. The mud is very mobile, and salinity is high. During low tide, water evaporates and salinity becomes higher than that of sea water. Under such conditions only marine species can survive, the commonest being the filamentous green alga, *Vaucheria.* This traps silt between its filaments, so raising the level of the mud. As it dies it yields humus, which improves the fertility of the mud. On some salt marshes the algae *Rhyizoclonium* or *Enteromorpha*, or the flowering plant *Zostera*, are found instead of *Vaucheria.*

In the richer and slightly raised mud the next colonizer is *Salicornia.* This is an annual plant, but its dead stems remain projecting from the mud and trap more silt. The decaying plants add more humus. *Salicornia* is the dominant plant of the landward part of the low salt marsh, but other plants such as *Plantago maritima* may also be found there. In the Solent the salicornia marsh has been followed by colonization by *Spartina townsendii*, which has spread rapidly, forming permanent spartina marsh. In other areas the succession proceeds differently from middle salt marsh.

2. *Middle salt marsh.* Among the *Salicornia* other plants become established. These include *Suaeda*, *Aster*, *Triglochin*, and *Puccinellia*. Usually, *Puccinellia* gradually becomes the dominant plant, and the *Salicornia*, being an annual plant, is unable to compete with the perennial plants that have now arrived. At this stage the soil level has risen and is covered by water only at the highest tides. Gradually other species become established, including *Armeria*, *Halimione*, *Limonium*, and *Spergularia*, but *Puccinellia* usually remains dominant. This stage of development of the salt marsh passes gradually to the final stage.

3. *High salt marsh.* This is covered only by the highest tides. The soil is waterlogged, but salinity is much lower than in the earlier salt marsh stages. *Puccinellia* is dominant, and plants of the middle salt marsh are common. Later *Juncus* and *Agropyron* become established and gradually replace the other species. Now the level of the soil has risen so that it is rarely flooded with brackish water, and the succession of the salt marsh is complete.

The next stage of the succession follows as the soil level builds up through decay of marsh plants and deposition of wind-blown soil. Conditions are hardly saline now, and plants that are not adapted to saline conditions can establish themselves. Frequently, grasses such as *Festuca rubra* and *Agrostis tenuis* enter at this stage, and the succession continues to *salt pasture* which can be used as grazing for farm stock.

FLORA OF SALT MARSHES

(for abbreviations, see endpapers)

DESCRIPTION	SCIENTIFIC NAME	COMMON NAME	M[1]	D[2]	HABIT
Flowering plants	*Agropyron pungens*	sea couch grass	7-9	C	R
	Armeria maritima	thrift, sea pink	4-10	C	W
	Aster tripolium	sea aster	7-10	C	b/S Y
	Glaux maritima	sea milkwort, black saltwort	6-8	C	S Y
	Halimione portulacoides	sea purslane	7-9	C	RW Y
	Juncus gerardii	mud rush	6-7	C	R
	Juncus maritimus	sea rush	7-8	C	S
	Limonum vulgare	sea lavender	7-10	C	W Y
	Plantago maritima	sea plantain	6-8	C	S
	Puccinellia maritima	sea poa	6-7	C	S
	Salicornia agg.	marsh samphire, glasswort	8-9	C	a Y
	Spartina × *townsendii*	cord grass	6-8	C	R
	Spergularia marina	lesser sea spurrey	6-8	C	a
	Spergularia media	greater sea spurrey	6-9	C	S
	Suaeda maritima	herbaceous seablite	7-10	C	a Y
	Triglochin maritima	sea arrow-grass	7-9	C	S
	Zostera spp.	eel-grass		C	
Algae	*Enteromorpha* spp.				
	Rhizoclonium spp.				
	Vaucheria spp.				FA

[1] Months of flowering [2] Distribution

Dunes

Salt marshes are formed by the deposition of water-borne silt, and in a similar way dunes are formed by the deposition of wind-borne sand. Deposition of silt occurs where water stagnates, and the same applies to the deposition of wind-borne sand. It is easy to observe that if any object, such as a stone or a fence or a picnic-basket, is on a windswept beach of dry sand the sand is deposited on the lee side of the obstruction, where wind speed is least. Given enough time, the deposit builds up until it is as high as the object. This is the beginning of the formation of a dune, but with an object of fixed size, such as a stone or a picnic basket, dune-building proceeds no further. If the object is a plant, a small dune forms on its lee side. The plant can grow, and as it grows more sand is deposited, and the dune grows, too. Few plants are able to initiate dunes because exposure to strong winds on the sea shore produces severe conditions. One species that is able to grow under these conditions and which can tolerate salt spray is *Agropyron junceiforme*. This is found on some coasts, producing small dunes close to the sea. The dunes grow, the plant gradually becomes partly buried, and the dunes can then grow no further. Further inland, away from the salt spray and gaining some protection from the seaward agropyron dunes, the marram grass, *Ammophila*, forms dunes in the same way as *Agropyron*. These dunes grow to much greater heights because *Ammophila* can form fresh roots farther up its stem, so it can 'climb' the dunes as they increase in height. *Ammophila* is thus responsible for most of the large dunes of our coasts, though *Carex arenaria* is the dune pioneer in some regions.

When first colonized by *Ammophila*, the dune soil, if such it can be called for it is almost pure sand, is very porous and dry. It rapidly becomes hot in sunshine and cools quickly at night. It is exposed to winds on the seaward side. On the lee side there is less exposure and, after a period of colonization by *Ammophila*, sufficient humus accumulates to allow other species to enter. These are a mixture of coastal plants, such as *Halimione* and *Carex arenaria* and species that are the early colonizers of hedgebanks (pp. 88-89) and garden beds (pp. 99-102). They include *Senecio jacobaea*, *Lotus corniculatus*, and the fescues (list of secondary colonizers, p. 170). They are mostly plants of universal distribution and exhibit the usual features of colonizers. Many are annuals and have rosettes of leaves close to the soil, or have low spreading stems.

In time, the dune soil, which has still risen owing to the deposition of sand between the primary and secondary colonizers, becomes completely covered with vegetation. Strong winds cannot now reach the surface, so that the dune cannot be eroded by wind. Neither can it be eroded by heavy rainfall. It has become a *fixed dune*. The gradual increase in humus has helped to make the sand more fertile, and improve its water retention. Many new species can enter (list for fixed dunes, pp. 170-171), among them being several leguminous species, well able to live in the nitrogen-deficient soil and helping to increase its nitrogen content. The damper soil conditions make it possible for mosses and lichens to grow as a ground layer beneath the herbaceous plants. The soil is not damp enough for the growth of liverworts.

The plants of fixed dunes may later become overshadowed by the growth of woody shrubs, such as *Rubus fruticosus*, *Rosa spinosissima*, *Lonicera*, *Sambucus*, *Crataegus*, and *Ligustrum*. Eventually, this succession may pass from scrub to the formation of woodland. In many areas fixed dunes are planted with pines, usually *Pinus nigra*.

Between dunes there are often low-lying marshy areas where pools of brackish water occur. These areas are called *dune slacks*; frequent colonizers are *Agrostis stolonifera*, *Plantago coronopus*, and *Bryum pendulum*. The water in slacks is rich in calcium from the broken shells of molluscs, so calcicolous plants are frequently found in dune slacks and also on dunes. Farther inland, behind the shelter of large dunes, conditions are more like those inland, and the slacks are colonized by many fenland species.

FLORA OF DUNES

(for abbreviations, see endpapers)

DESCRIPTION	SCIENTIFIC NAME	COMMON NAME	M[1]	D[2]	HABIT
Primary colonizers	*Agropyron junceiforme*	sand couch grass	6-8	C	R
	Ammophila arenaria	marram grass	7-8	C	XR
	Carex arenaria	sand sedge	6-7	C	S
	Honkenya peploides		5-8	C	S
Secondary colonizers	*Calystegia soldanella*	sea bindweed	6-8	C	S
	Carex arenaria	sand sedge	6-7	C	S
	Erodium maritimum	sea storksbill	5-9		a
	Festuca ovina	sheep's fescue	5-8	U	S
	Festuca rubra	red fescue	5-7	U	S
	Halimione portulacoides	sea purslane	7-9	C	RWY
	Hypochaeris radicata	cat's ear	6-9	U	S
	Lotus corniculatus	bird's-foot trefoil, eggs-and-bacon	6-9	U	S
	Senecio jacobaea	ragwort	6-10	U	b/S
Fixed dunes	*Asperula cynanchica*	squinancy wort	6-7	+	S
	Carex arenaria	sand sedge	6-7	C	S
	Carlina vulgaris	carline thistle	7-10	S+	T J
	Cerastium semidecandrum	little mouse-ear chickweed	4-5		a
	Erophila verna	spring whitlow-grass	3-6		aJ
	Eryngium maritimus	sea holly	7-8	C	S
	Festuca rubra	red fescue	5-7	U	S
	Hieracium pilosella	mouse-ear hawkweed		U	R
	Holcus lanatus	Yorkshire fog	6-9	U	S
	Lotus corniculatus	birds'-foot trefoil, eggs-and-bacon	6-9	U	S
	Myosotis ramosissima	early forget-me-not	4-6		a
	Ononis spinosa	restharrow	6-9		W Z

[1] Months of flowering [2] Distribution

DESCRIPTION	SCIENTIFIC NAME	COMMON NAME	M	D	HABIT
	Phleum arenarium	sand cat's-tail	5-6	C	a
	Potentilla anserina	silverweed	6-8	U	S
	Potentilla reptans	creeping cinquefoil	6-9	S	S
	Sedum acre	wall-pepper, biting stonecrop	6-7		S
	Thymus drucei	wild thyme	5-8		S
	Trifolium arvense	hare's-foot	6-9		a
	Trifolium pratense	red clover	5-9	U	S
	Trifolium repens	white clover	6-9	U	S
Mosses of fixed dunes	*Brachythecium albicans*				FM
	Bryum spp.	thread mosses			CM
	Ceratodon purpureus	purple-fruiting heath moss			CM
	Tortula ruraliformis	screw moss			CM
Lichens of fixed dunes	*Cladonia* spp.				LL
	Evernia prunastri				BL
	Parmelia spp.				LL
	Peltigera spp.				LL

Cliffs

Cliffs are formed by erosion and weathering, which continually expose new rock. The kinds of plants and animals found on cliffs are thus very dependent on the nature of the rock, and the way in which it weathers or becomes eroded. Granite and other ancient rocks weather slowly, and form little soil. The cliffs consist almost entirely of steep rock faces to which few plants other than mosses, lichens, and algae can attach themselves. Only on ledges and in crevices will sufficient soil eventually accumulate to allow higher plants to grow. On parts of our coasts, cliffs are of chalk or of limestone. This weathers more freely and the cliff tops and ledges are more quickly covered by soil in which plants can grow. These include many calcicolous species. Where the land consists of sandy or pebbly deposits the cliffs have a crumbling and less steeply-sloping face. Plants can more easily take root on this surface, which is already soil-like, though it contains relatively little humus and is subject to washing away and sliding. The situation is similar to that of the hedgebank. New areas of soil are constantly being exposed by landslides and by wind or rain erosion, and here annual plants, with rosette leaves or creeping habit, are the first to colonize. In the more stable areas these may be followed by perennial plants, eventually including woody perennials such as *Sambucus*, *Crataegus*, and *Sorbus aucuparia*.

The vegetation often varies according to the height above the beach. At the base of the cliff one usually finds robust plants with leathery leaves or plants forming dense cushions. Such plants are *Silene*, *Armeria*, and *Beta*. These plants are resistant to the salt spray that occurs in the region of the cliff base.

Nearer the cliff top there is no salt spray, though fine particles of salt from spray that has evaporated can be carried up by air currents. Cliff-top conditions are difficult for plants owing to the exposure to sun and wind. Drainage is easy down the steep cliff face so that soils are dry. Under these conditions we would again expect to find plants with the rosette, cushion or mat-forming habit. Examples of these found on or near cliff-tops are *Thymus*, *Helianthemum*, *Silene*, *Cochlearia*, *Plantago coronopus*, *Erodium* and *Sagina*. Most of these are plants especially characteristic of coastal habitats, but other plants of waste land or meadows or of grassland may be found, especially where exposure is less extreme.

The farther one gets from the cliff edge, the more conditions become like those of inland habitats. Exposure to wind and salt and sunshine becomes progressively less. Within a few miles of the coast inland species become commoner, but in this zone one often finds certain species that are not necessarily cliff or shore dwellers but which seem to do best in areas near to coasts. Examples are *Apium graveolens* (wild celery), *Carduus tenuiflorus* (slender thistle), *Foeniculum vulgare* (fennel), and *Smyrnium olusatrum* (Alexanders). The common inland tree *Acer pseudoplatanus* (sycamore) also does well in coastal regions.

Animal life on cliffs depends much on what vegetation cover there is. If this is adequate, a wide range of small invertebrate species may be found, particularly insects, spiders, centipedes, and millipedes. Compared with other habitats, cliffs have many inaccessible spots, which makes them an ideal place for birds. They can gain access to ledges, sheer rock faces and peaks denied to terrestrial predators. Though the cliff affords excellent protection it provides little food. Some coastal birds (list p. 174) feed on the shore, but others feed out at sea or inland.

FLORA OF CLIFFS

(for abbreviations, see endpapers)

DSECRIPTION	SCIENTIFIC NAME	COMMON NAME	M[1]	D[2]	HABIT
	Armeria maritima	thrift, sea pink	4-10	C	W
	Beta vulgaris ssp. *maritima*	sea beet	7-9	C	b
	Cochlearia officinalis ssp. *officinalis*	common scurvy-grass	5-8	C	b/T
	Erodium maritimum	sea storksbill	5-9		a
	Helianthemum chamaecistus	common rockrose	6-9	+	W
	Plantago coronopus	buck's-horn plantain	5-7	C	S
	Sagina maritima	sea pearlwort	5-9	C	a
	Silene maritima	sea campion	6-8	C	SY
	Spergularia rupicola	cliff sea spurrey	6-9	C	WS
	Thymus drucei	wild thyme	5-8		S

[1] Months of flowering [2] Distribution

BIRDS found on coasts

SCIENTIFIC NAME	COMMON NAME	NEST SITE
Anthus spinoletta	rock pipit	cliffs, rocks
Columba livia	rock dove	ledges, caves
Corvus corone	carrion crow	ledges
Falco tinnunculus	kestrel	ledges
Fratercula arctica	puffin	burrows in turf on cliff top
Larus argentatus	herring gull	ledges, etc.
Larus canus	common gull	low grassy slopes
Larus fuscus	lesser black-backed gull	grass-covered islands
Larus marinus	great black-backed gull	any inaccessible sites
Phalacrocorax aristotelis	shag	rocky coasts, caves or ledges
Phalacrocorax carbo	cormorant	ledges
Pyrrhocorax pyrrhocorax	chough	cliff holes
Riparia riparia	sand martin	cliff holes
Sterna hirundo	common tern	bare rocks, shingle beds, mud flats
Sula bassana	gannet	ledges
Tringa totanus	redshank	anywhere
Uria aalge	guillemot	ledges

BOOK LISTS

The first list contains key works, books used for identification and usually containing keys that are an essential part of the equipment of the field biologist. Those most suitable and useful for the beginner are marked *.

Other books useful in identification, but not containing keys, are included in the list on pp. 176-178.

† Paperback: Fontana New Naturalist Series.
O/P: out of print books which may, however, still be available in libraries.

1* BARRETT, J. H. and YONGE, C. M., *Pocket Guide to the Seashore* (Collins); £1.50.

2* BRIGHTMAN, F. H. and NICHOLSON, B. E., *Oxford Book of Flowerless Plants* (Oxford University Press); £2.25.

3 CHANCELLOR, R. J., *The Identification of Weed Seedlings of Farm and Garden* (Blackwell, Oxford); 75p.

4 CHRYSTAL, R. N., *Insects of the British Woodlands* (Warne); 85p.

5 CLAPHAM, A. R., TUTIN, T. G. and WARBURG, E. F., *Flora of the British Isles* (Cambridge University Press). The complete flora; rather too expensive to be recommended for work at *junior* secondary level; £4.

6* CLAPHAM, A. R., TUTIN, T. G. and WARBURG, E. F., *Excursion Flora of the British Isles* (Cambridge University Press). Shorter and cheaper than the above and adequate on nearly all occasions. Recommended; £1.50.
Four books of illustrations to the above floras are published, drawn by SYBIL J. ROLES; £2.25 each.

7* CLEGG, J., *The Observer's Book of Pond Life* (Warne); 30p.

8* CLOUDSLEY-THOMPSON, J. L. and SANKEY, J., *Land Invertebrates* (Methuen); £1.05.

9 COLYER, C. N. and HAMMOND, C. D., *Flies of the British Isles* (Warne); £2.75.

10* CORBET, G. B., *The Identification of British Mammals* (British Museum (NH)); 20p.

11 DAHL, E., *Analytical Keys to the British Macrolichens* (Cambridge). Cyclostyled copies available from the Secretary, The British Lichen Society, Department of Botany, British Museum (NH), Cromwell Road, London S.W.7. Unfortunately, this key contains few of the species found growing on walls; 40p.

12 DALE, A., *Patterns of Life* (Heinemann). A general work, but the keys are numerous and good, so this *is* worth buying for the keys alone; 62p.

13* DICKINSON, C. I., *British Seaweeds* (Eyre and Spottiswoode); £1.25.

14* ELLIS, C., *The Pebbles on the Beach* (Faber and Faber); hardback £1.05, paperback 45p.

15* FITTER, R. S. R. and RICHARDSON, R. A., *Pocket Guide to British Birds* (Collins); £1.50.

16* FITTER, R. S. R. and RICHARDSON, R.A., *Pocket Guide to Nests and Eggs* (Collins); £1.25.
Some may prefer the alternative PETERSON, MOUNTFORD and BOLLOM, *Field Guide to the Birds of Britain and Europe* (Collins); £2.00.

17 FRESHWATER BIOLOGICAL ASSOCIATION. Various keys to freshwater animals. List obtainable from Freshwater Biological Association, The Ferry House, Ambleside, Westmorland.

18 HELLYER, A. G. L., *Garden Pests and Diseases* (W. H. and L. Collingridge, London); £1.05.

19 HUBBARD, C. E., *Grasses* (Penguin); 50p.

20 JACKSON, A. B., *The Identification of Conifers* (Arnold); 62p.

21 JEREMEY, A. C. and TUTIN, T. G., *British Sedges* (Botanical Society of the British Isles); £1.05.

22* LANGE, M., and HORA, F. B., *Guide to Mushrooms and Toadstools* (Collins); £1.50.

23 LAURENCE, M. J. and BROWN, R. W., *Mammals of Britain; their tracks, trails and signs* (Blackwell); £1.50.

24 LEWIS, T. and TAYLOR, L. R., *Introduction to Experimental Ecology* (Academic Press); £1.88.

25* MACAN, T. T., *A Guide to Freshwater Invertebrate Animals* (Longman); 60p.

26* MCLINTOCK, D. and FITTER, R. S. R., *Pocket Guide to Wild Flowers* (Collins); £1.25.

27 MEIKLE, R. D., *British Trees and Shrubs* (Eyre and Spottiswoode); £1.25.

28* MILES and MILES, *Chalkland and Moorland Ecology* (Hulton); 75p.

29* NUFFIELD FOUNDATION, *Keys to Small Organisms in Soil, Litter and Water Troughs* (Longman/Penguin); 15p.

30* PAVIOUR-SMITH, K. and WHITTAKER, J. B., *A Key to the Major Groups of British Free-living Terrestrial Invertebrates* (Blackwell, Oxford); 13p.

31* PRIME, C. T. and DEACOCK, R. J., *Trees and Shrubs* (Heffer, Cambridge); 75p.

32 SANKEY, J., *A Guide to Field Biology* (Longman). Has a key for soils; 85p.

33 SOUTHWOOD, T. R. E. and LESTON, D., *Land and Water Bugs of the British Isles* (Warne); £1.85.

34 SOUTHERN, H. N. (Ed.), *The Handbook of British Mammals* (Blackwell, Oxford); £2.25.

35 TAYLOR, P. G., *British Ferns and Mosses* (Eyre and Spottiswoode); £1.25.

36 TEBBLE, N., *British Bivalve Shells* (British Museum (NH)); hardback £1, paperback 70p.

37* VEDEL, H. and LANGE, J., *Trees and Bushes in Wood and Hedgerow* (Methuen); 90p.

38 WATSON, E. V., *British Mosses and Liverworts* (Cambridge University Press); £4.

39 KERRICH, G. J., MEIKLE, R. D. and TEBBLE, N., *Bibliography of Key Works for the Identification of the British Fauna and Flora* (The Systematics Association; obtainable from The Treasurer, The Systematics Association, c/o The British Museum (NH), Cromwell Road, London S.W.7); £1.

The second list contains books for reference for background reading, and for further study of particular communities. Those specially recommended are marked *.

1 ARY, S. and GREGORY, M., *Oxford Book of Wild Flowers* (Oxford University Press); £1.75.

2* ASHBY, M., *Introduction to Plant Ecology* (Macmillan). Some ideas for investigations; £2.25.

3* BENNETT, D. P. and HUMPHRIES, D. A., *Introduction to Field Biology* (Arnold); 90p.

4 BISHOP, O. N., *Outdoor Biology* (Murray); Books 1-3, 50p each; Teachers' Guide, 80p.
5 BROWN, E. S., *Life in Fresh Water* (Oxford University Press); O/P.
6 BURTON, J., YARROW, I. H. H., ALEN, A. A., PARMENTER, L. and LANSBURY, J., *Oxford Book of Insects* (Oxford University Press); £2.50.
7 CAULTON, E., *An Ecological Approach to Biology* (*J. Biol. Educ.* 1970, **4**, 1, 1-10).
8 CLOUDSLEY-THOMPSON, J. L., *Microecology* (Arnold); 38p.
9 DARLINGTON, A., *Ecology of Refuse Tips* (Heinemann); £1.25.
10 DARLINGTON, A., *Natural History Atlas of the British Isles* (Warne); £1.75.
11 DARLINGTON, A., *The Pocket Encyclopaedia of Plant Galls* (Blandford); £1.25.
12 DRABBLE, H., *Plant Ecology* (Arnold); O/P.
13 ELLIS, E. A., *The Broads* (Collins); £1.50.
14 ENNION, E. A. R. and TINBERGEN, N., *Tracks* (Oxford University Press); £1.40.
15 FITTER, R. and FITTER, M., *The Penguin Dictionary of British Natural History* (Penguin); 42p.
16 FRIEDLANDER, C. P., *Heathland Ecology* (Heinemann); 52p.
17 GILMOUR, J. and WALTERS, S. M., *Wild Flowers* (Collins); £1.50.
18 HELLYER, A. G. L., *Garden Pests and Diseases* (W. H. and L. Collingridge, London); £1.05.
19 JENKINS, T., *Fishes of the British Isles* (Warne); £1.05.
20 JEWELL, A. L., *Observer's Book of Mosses and Liverworts* (Warne); 30p.
21 JONES, F. G. W. and JONES, M., *Pests of Field Crops* (Arnold); £3.
22 KEVAN, D. K. MCE., *Soil Animals* (Witherby). Useful for soil projects; £1.75.
23* KNIGHT, M., *Frogs, Toads and Newts in Britain* (Brockhampton); 62p.
24 LEUTSCHER, A., *A Study of Reptiles and Amphibians* (Blandford); 52p.
25 LEUTSCHER, A., *Field Natural History* (Bell); £3.
26 LEUTSCHER, A., *Tracks and Signs of British Animals* (Cleaver-Hume); £1.05.
27 LEWIS, J. R., *The Ecology of Rocky Shores* (English Universities Press); £2.10.
28* LEWIS, T. and TAYLOR, L. R., *Introduction to Experimental Ecology* (Academic Press); £1.88.
29 LYNEBORG, L., *Field and Meadow Life* (Blandford); 68p.
30* MACAN, T. T., *Freshwater Ecology* (Longman); £2.25.
31 MACAN, T. T. and WORTHINGTON, E. B., *Life in Lakes and Rivers* (Collins); £1.80.
32 MANDAHL-BARTH, G., *Woodland Life* (Blandford); 68p.
33* MARTIN, W. K., *The Concise British Flora in Colour* (Ebury Press and Michael Joseph); £2.25.
34 MATTHEWS, L. H., *British Mammals* (Collins); £1.75.
35* MELLANBY, H., *Animal Life in Fresh Water* (Warne); £1.25.
36 NATURE CONSERVANCY, *Nature Trails* (Warne); 25p.
37 NEAL, E. G., *Woodland Ecology* (Heinemann); 52p.
38 PEARSALL, W. H., *Mountains and Moorland* (Collins); £1.50 (67p †).
39 PHILLIPSON, J., *Ecological Energetics* (Arnold); 38p.
40 POPHAM, E. J., *Some Aspects of Life in Fresh Water* (Heinemann); O/P.
41 PRUD'HOMME VAN REINE, W. J., *Plants and Animals of Pond and Stream* (Murray); 75p.
42 PRUD'HOMME VAN REINE, W. J., *Plants and Animals of the Seashore* (Murray). O/P.
43 RUSSELL, SIR E. J., *The World of the Soil* (Collins); £1.15 (47p †).
44 SALISBURY, SIR E., *Weeds and Aliens* (Collins); £1.50.
45* SANKEY, J., *A Guide to Field Biology* (Longman); 75p.
46 SANKEY, J., *Chalkland Ecology* (Heinemann); 80p.
47 SOUTHERN, H. N. (Ed.), *The Handbook of British Mammals* (Blackwell, Oxford); £2.25.
48* SOUTHWARD, A. J., *Life on the Seashore* (Heinemann); 80p.
49 STAMP, L. D., *Britain's Structure and Scenery* (Collins); £1.50 (47p †).

50 STOKOE, W. J., *Observer's Book of Grasses, Sedges and Rushes* (Warne); 30p.

51* TANSLEY, SIR A. G., *Britain's Green Mantle* (Allen and Unwin); £1.75.

52 TANSLEY, SIR A. G., *The British Isles and their Vegetation* (Cambridge University Press); £8 set of 2 vols.

53 YONGE, C. M., *The Seashore* (Collins); £1.50 (67p †).

54* ZIM, H. S. and SHAFFER, P. R., *Rocks and Minerals* (Hamlyn). This and others of the Little Guides series (birds, fossils, wild flowers) are cheap and well illustrated in full colour; 25p.

INDEX

Species are not included in the index. The reader is referred to the flora and fauna lists, where species are listed in alphabetical order under various groupings.

Index entry prefixed by P indicates a reference to a photographic plate.